Vibration Problems in Machines

in Machines

Diagnosis and Resolution

Vibration Problems in Machines

Diagnosis and Resolution

ARTHUR W. LEES

CRC Press
Taylor & Francis Group
Boca Raton London New York

CRC Press is an imprint of the
Taylor & Francis Group, an **informa** business

CRC Press
Taylor & Francis Group
6000 Broken Sound Parkway NW, Suite 300
Boca Raton, FL 33487-2742

First issued in paperback 2017

© 2016 by Taylor & Francis Group, LLC
CRC Press is an imprint of Taylor & Francis Group, an Informa business

No claim to original U.S. Government works

ISBN-13: 978-1-4987-2674-0 (hbk)
ISBN-13: 978-1-138-89383-2 (pbk)

Visit the Taylor & Francis Web site at
http://www.taylorandfrancis.com

and the CRC Press Web site at
http://www.crcpress.com

Contents

Preface

The understanding of vibration signals from turbo-machinery is an important feature of many industries. It forms the basis of condition monitoring and is of crucial importance in the rapid diagnosis and rectification of faults. The field calls for a variety of skills ranging from instrumentation expertise to mathematical manipulation and signal processing. The focus is on gaining a physical understanding of the processes giving rise to vibration signals. This may be viewed as a process of seeking to infer internal conditions from (sometimes limited) external measured data.

This often requires some form of mathematical model, and the approach taken is to use some basic finite element analysis to study system behaviour. These methods were first applied to rotating machinery several decades ago, but with progress in computing and the advent of packages such as MATLAB®, the ease with which the concepts may be applied has dramatically improved. It is now a relatively straightforward matter to examine proposed design changes – in effect, to perform numerical experiments.

The material in this book stems both from my experiences in the industry and, in more recent years, of teaching and research in academia. Inevitably, my own experiences have, to some degree, influenced my choice of examples. In my various posts, I have had fruitful interactions with many people who have significantly enhanced my perception of problems in machinery. There are too many people involved to thank individually, but I must mention three. Professor Mike Friswell and I have worked together closely for 20 years, and I acknowledge innumerable helpful and illuminating discussions. I also thank Professor John Penny for his help with proofreading and a number of helpful suggestions on the structure of the text. Both of these people were my co-authors, together with Professor Seamus Garvey, of the book I refer to in many places which gives a more complete discussion of rotor dynamics. My deepest thanks go to my wife, Rita: Not only has she given me grammatical advice, but has also helped maintain my sanity throughout the final months of preparing this text.

Arthur W. Lees

Author

Professor Arthur W. Lees, BSc, PhD, DSc, CEng, CPhys, FIMechE, FInstP, LRPS, graduated in physics and remained in Manchester University, Manchester, United Kingdom for three years' research. After completing his PhD, he joined the Central Electricity Generating Board, London, United Kingdom, initially developing finite element codes and later resolving plant problems.

After a sequence of positions, he was appointed head of the Turbine Group for Nuclear Electric Plc, London, United Kingdom. He moved to Swansea University in 1995 and has been active in both research and teaching.

He is a regular reviewer of many technical journals and was, until his recent retirement, on the editorial boards of the *Journal of Sound and Vibration* and *Communications in Numerical Methods in Engineering*. His research interests include structural dynamics, rotor dynamics, inverse problems and heat transfer. Professor Lees is a fellow of the Institution of Mechanical Engineers and the Institute of Physics, a chartered engineer and a chartered physicist. He was a member of the Council of the Institute of Physics, 2001–2005. He is now professor emeritus at Swansea University but remains an active researcher.

MATLAB®

Modelling can be carried out using the package of MATLAB scripts freely available at www.rotordynamics.info. A further toolbox is under construction for the following studies:

a. Rigid rotor analysis
b. Flexible rotor analysis
c. Single plane balancing
d. Two plane balancing
e. Modal balancing
f. Gear torsional analysis
g. Transient predictions
h. Catenary calculations

This toolbox and additional electronic products available at the CRC Press website: http://www.crcpress.com/product/isbn/9781498726740.

MATLAB® is a registered trademark of The MathWorks, Inc. For product information, please contact:

The MathWorks, Inc.
3 Apple Hill Drive
Natick, MA 01760-2098 USA
Tel: 508-647-7000
Fax: 508-647-7001
E-mail: info@mathworks.com
Web: www.mathworks.com

1

Introduction

The general field of condition monitoring has received substantial attention over the last few decades and it is worth reflecting on the state of the topic, because, although it has always been practised at some level, the manner in which condition is assessed is constantly under review in the light of recent developments in understanding.

In assessing the condition of a piece of equipment, the operator gathers data such as vibration, operating temperature, noise, performance and electrical parameters where appropriate. At one time, the comparison with normal condition was achieved largely on the basis of staff experience, but the general trend has been towards a more precise quantified approach. This has been required by revised patterns of working and increasing plant complexity, but it is, in essence, the same operation. A fundamental question arises of how one can 'codify' the knowledge of an experienced engineer and focus the knowledge on a specific area of plant. This presents indeed a challenge which has shown significant progress in recent years, although the issue cannot be regarded as completely resolved. Important progress has been made in computational modelling (both finite element analysis (FEA) and computational fluid dynamics (CFD)), artificial neural networks (ANNs), statistical approaches, expert systems and identification methods. All of these have a role to play in assessing the condition of a piece of equipment and their role will be outlined in subsequent chapters. First of all however, the general field of condition monitoring, as applied to rotating machines, is reviewed.

1.1 Monitoring and Diagnosis

Generally, these two terms are linked under the general heading of condition monitoring, but in fact, there are two quite distinct functions. In both areas, the first requirement is to gather and record all salient details of the operation of the piece of equipment but, as will be discussed, the choice as to what details are salient is a far from trivial task. However, that discussion is deferred for the present.

To illustrate this point with a specific example, let us consider a centrifugal pump driven by an electric motor. In such a case, the monitored parameters would include bearing vibration levels, temperature, water pressure, water flow rate, motor current and voltage. Note that although this is a fairly long list, it is by no means exhaustive. In some circumstances, one may wish to record the rotor vibration (as opposed to that of the bearings) and bearing oil temperature. In fact, even a relatively simple piece of machinery may have a significant number of parameters which may be useful for monitoring purposes and a judicious choice is required to limit the measured set to cost-effective proportions; however, making this choice requires some appreciable physical insight.

Having decided on a set of monitored parameters, some method of recording is the next choice to be made and this ranges from regular spot checks to some form of continuous monitoring, now almost invariably computer based. Whilst the latter represents a more expensive option, it does offer more flexibility in terms of the ways in which data can be manipulated to offer insight into the underlying features of machine operation. Here again decisions are required which demand physical insight into machine operation and the likely failure scenarios.

We now consider some of the ways in which plant data may be analysed and how this may be used to form judgements about plant operation. Clearly, any general trend in the plant data, or indeed a sudden change, suggests that the equipment has changed in some way and, subject to some checks, may require the removal of the plant from service. Note that realistically, all equipment is subject to some random perturbations and so, statistical techniques are needed to form a valid decision such as when to remove plant from service.

To proceed further, we examine the three basic questions that are posed in condition monitoring systems which are

1. Has something gone wrong?
2. What has gone wrong?
3. How long can the plant run safely?

To address the first of these questions, it is often sufficient to adopt a purely statistical analysis of monitored data and seek trends and changes. At the most basic level, no knowledge of the internal operation of the machinery is required. For example a monitoring system may simply plot the overall vibration levels at the bearings (or elsewhere) and not examine the (short-term) time variation. Some of the ways in which data can be examined is discussed in Chapter 2. As discussed in later chapters, a great deal depends on the frequency with which measurements are monitored and provided several measurements are recorded during each rotor revolution, the orbit of the shaft may be traced and this gives some further insight into the machine's behaviour. In many cases, rudimentary measurements will give some indication that something has happened to a machine and straightforward comparison with records will suffice to answer the first of the three questions. To answer the second, however, generally requires considerably more insight which may be provided by extensive experience, a detailed theoretical analysis or, very often, a combination of these. The third question, as to how long the plant will/can run is more difficult still, and is to a large extent still at the research stage. Nevertheless, an in-depth understanding of the machine's operation is an essential pre-requisite. In later chapters, a discussion is given as to the interpretation of plant data, but first, we give a brief survey of quantities that may be useful in assessing a machine's behaviour.

1.1.1 Monitored Parameters

1.1.1.1 Vibration

On rotating plant, vibration is the most commonly monitored parameter. The reasons for this are twofold: first, it is readily measured with convenient instrumentation and, perhaps more important, vibrations give a comprehensive reflection of the state of a rotating machine. The disadvantage is that much of the information for diagnosis can only be gained after extensive data processing, but standard techniques have now been developed which go some way towards relieving this burden. For basic monitoring, vibration levels can be, and are, used to great effect.

1.1.1.2 Pressure, Flow and Temperature

In the case of pumps, pressure and flow represent the main performance parameters, and hence, it is important to monitor these on a regular basis, although clearly, these will not be expected to vary as rapidly as vibration. There may, of course, be fluctuations in pressure, but flow rate will not track these owing to substantial inertia effects. Chapter 6 has a discussion of some of the ways in which the pressure field within a typical centrifugal pump has a direct influence on the vibration characteristics. The treatment of this is by no means comprehensive, and the interested reader is referred to the work of Childs (1993). The important point to emphasise here is that various pieces of information are inter-related and, taken together, they present a complex picture; it is the role of the diagnostic engineer to form an overall view from this complex pattern.

1.1.1.3 Voltage and Current

Electrical measurements also form an important part of this picture. A very large number of machines are driven by electric motors and the electrical measurements can be used to yield information on the overall machine efficiency, a key indicator of general deterioration. Used at this level, periodic checks would suffice, but more detailed analysis of fluctuations gives added information on both the motor and the auxiliary machine which is being driven.

1.1.1.4 Acoustic Emission

Acoustic emission (AE) is, in one sense, simply vibration at very high frequency, but this definition fails to give due recognition to its important distinction from conventional vibration data. As its name suggests in looking at AE, the engineer is monitoring the acoustic energy emitted by the material as changes (e.g. crack formation and propagation) take place. In effect one is 'listening' to the high-frequency waves generated by the breaking of inter-molecular bonds within a component, rather than the direct consequences of externally applied forces. Until very recent times, AE was used rather as a course monitor to count so-called events, as a guide to the presence or otherwise of cracking activity. For many years, it has been used to monitor the integrity of structures, but it is only much more recently (Price et al. 2005, Sikorska and Mba 2008) that it has been applied to rotating machinery. Improving hardware and software has facilitated the examination of the frequency resolution of AE signals, and this too has enhanced our understanding. Measurements are often made in the range of several hundred kHz as this is reasonably easy to obtain and process with modern equipment and, it turns out, yields information of interest. A common misunderstanding is that these frequencies are characteristic of the breaking bonds within the material, but this is not the case: such bonds give frequencies which are several orders of magnitude higher. The more accurate picture is that a breaking bond generates a very short pulse which contains a wide range of frequencies. As the waves travel, those which resonate within some part of the body become predominant and hence, the spectrum of the AE may be used to tell the operator the nature of the body around the fault. It is the essence of AE that it provides a good approach to the identification of highly localised phenomena.

1.1.2 Fault Localisation

The effective use of condition monitoring and condition-based maintenance encompasses a whole hierarchy of inter-related disciplines. At what might be termed level 1, purely statistical approaches can be applied to measured data to detect if there has been some underlying change to the condition of the machine, and some insight into such approaches is given in Chapter 2. This type of analysis can be carried without any knowledge of the machine's construction or operation and this is both a strength and a weakness: it means that the operator can determine if a fault/change has occurred without any assumptions as to the physical processes, but nothing further can be gleaned about the location or nature of the fault.

It may appear at first sight that the requirement to localise a fault is largely academic, but this is far from the case. On large machines, such as turbo-alternators, many days of production can be saved by focussing maintenance work on the appropriate part of the machine. However, to make any valid progress on localisation requires insight or prior knowledge into the machine's design and dynamic properties. Foremost among the requirements is a knowledge of the machines' natural frequencies (critical speeds) and mode shapes. This knowledge may be developed by operational experience, plant tests and/or validated mathematical models, most commonly finite element (FE) models. More details on the requirements of these models are given in Section 1.2. The essential point at this stage is to emphasize that although condition monitoring is carried out without any reference to models for some simple machines, the potential is greatly extended by the use of models. This is because their key role is to relate the external measurements to the internal operating conditions.

Having established a model, the determination of natural frequencies and mode shapes is a straightforward matter and is discussed in many text books (see, e.g. Friswell et al. 2010, Inman 2008). The mode shapes will often give some clues as to the important locations for particular types of fault. Conclusive location identification may require some further analysis.

1.1.3 Root Cause

The issue of root cause is a theme running throughout this text and embraces the issues of fault localisation and frequency composition of the vibration signals. Chapters 4 and 5 discuss a range of common machine faults and the types of vibration signal they give rise to. The process of fault identification is basically one of recognising the appropriate patterns of response and matching them to fault types. More importantly, the process may be seen as one of gaining insight into the physical processes within the machine.

One might imagine that it would be easy to automate this process, but whilst there has been progress, there is still the need for a human expert on more complex machines. Recent research in this area has focussed on two areas: the development of extended models as discussed in Chapter 7 and the development of expert systems as in Chapter 8. For the most complex machines however, no fully automated system is likely to be available in the foreseeable future.

1.1.4 Remaining Life

Whilst the issues of fault localisation and root cause raise some problems, the identification of remaining life remains the 'Holy Grail of all Condition Monitoring'. It is an extremely complex topic and involves input from a number of disciplines. Whilst in some instances, reasonable predictions can be made, in most cases, these rely substantially on practical experience of the plant involved. Predictions can be made on, for example, crack growth rates, but these rely heavily on materials data which are in some instances subject to substantial error.

1.2 Mathematical Models

Many of the principles throughout this text are explained and illustrated with mathematical models, in some cases following analytic expressions and in others a numerical model. Some readers may find this surprising as in many cases, a machine operator may not have precise data with which to formulate such models: this will be so particularly in specifying fine clearances which will change with time as the machine components wear, but this does not invalidate the need or the role for theoretical models – in many ways, it emphasises the role. The aim of the model is to enhance and extend operational experience: to a degree, the actual numbers are secondary and what really matters is the extent to which the different parameters influence each other. This type of sensitivity is extremely valuable in gaining an understanding of the machine's operation. The sentiment which may surprise some readers is completely aligned with the theme presented here; although mathematical models and computing (and hence numbers) are used extensively throughout the text, the basic motivation is to elicit understanding of the underlying physical processes as opposed to any precise prediction. It is far more important to use models to examine interdependencies of various parameters and to understand the way in which physical changes (or uncertainties) can/will influence measured behaviour. Whilst the onset of problems can often be identified using purely statistical or empirical approaches, a full diagnosis and localisation of the fault will almost invariably require some form of model. This should not be surprising as the model is simply a means of expressing knowledge of the operation of a system.

Such an approach will not always be appropriate. For small items of plant which are readily interchangeable, a suitable approach would be to remove the item from service pending a general overhaul. For high capital plant, however, this approach is rarely feasible and it is in such cases that fault diagnosis, as a means to minimising maintenance/repair time, is imperative: this is where it becomes vital to glean all the information possible from the data available.

The need to gain insight into a machine's operational behaviour raises another interesting contrast to which we will return several times, namely, the distinction between data and information. This distinction is discussed more quantitatively in Chapter 7 in the context of elucidating the influence of foundations on the dynamics of machines, an effect which is particularly important with very large machines such as turbo-generators.

For a given machine, all these parameters will be linked and one of the aims of mathematical modelling must be to elucidate these links with a view to understanding the fluctuations which occur in machines running under nominally steady-state conditions. To some extent, steady state is a figment of the engineer's imagination: a machine may be operating at a steady (average) speed and load, but in reality, there will always be some short-term fluctuations: understanding these variations will often yield significant insight into a machine's operation. Examples of this are given in connection with turbo-generator operation in Chapter 4 and gearbox oscillations on a pumping system are discussed in Chapter 5, together with a detailed case study in Chapter 9. In many cases, a full understanding of the connections between variables is beyond our current analysis methods, principally due to a lack of detailed knowledge of machine parameters such as subsoil properties beneath the machines. Indeed, there is ample statistical evidence that at coastal power plants, the state of the tide has a substantial influence on vibration levels of machines. Whilst the mechanism of this interaction is understood, modelling techniques have not reached the level of quantitative predictions of such effects. In such cases, ANNs can yield insight by effectively forming a non-linear curve fit to the measured data. Chapter 8 gives some examples of this type of study: such approaches, whilst not an acceptable end in themselves, may well aid progress towards more complete models.

As focus changes to the second question, the diagnosis stage, significantly more detail of data (or, at least the analysis of that data), together with some input of the physical processes of the machine's operation, is essential. A prime requirement for any detailed study is the time development of the signal. (Here we refer primarily to vibration, pressure or voltage signals: temperature variations will normally be very much slower.) Often these signals will be analysed in terms of frequency and, in particular, their relationship with the shaft rotation speed. Usually this relationship will give a good indicator as to the nature of the problem and will be the first stage in the diagnosis. Although some small items of plant may not warrant a thorough diagnosis of any fault, the situation on a high capital item, such as a turbo-alternator, is very different. The cost, in terms of lost revenue of one of these large machines, can be up £0.5 M per day. Because of this high revenue cost, it is very important to minimise the time required in rectifying the fault and return the unit to service, and this means that fault diagnosis and localisation can be extremely cost-effective. Some of the techniques available for this important task are reviewed in subsequent chapters.

The third question posed relating to plant lifetime is rather more difficult and is, in general, still at the research and development stage: nevertheless, useful predictions can be made with care. This can be extremely beneficial commercially, for instance, by deferring an emergency outage and completing necessary repairs within a scheduled maintenance period. To achieve this requires a combination of understanding the fault, predicting its likely progress, modelling the influence on the machine's behaviour and carefully tracking this against monitored parameters until the scheduled removal from service.

1.3 Machine Classification

The range of rotating machines is immense and any attempt at categorisation may appear at first sight daunting. Rotor masses span over a range of about eight orders of magnitude. Whilst small machines may have a rotor with a mass of a few grams, some large machines have rotating elements weighing several hundred thousand kilograms. Similarly, the range of speed is immense: whilst wind turbines typically rotate at around 1/3 revolutions per second (rev/s), turbo-molecular pumps run around 120,000 rev/s and micro-turbines currently being developed operate at over 1 million rev/s.

Given this wide variation, classification of rotating machine is not a simple matter, but it is necessary to progress with the basic task of this book, namely, the elucidation and clarification of some of the main techniques for the interpretation of vibration signatures. Some of the key determining parameters are mass of rotor, load on bearings, rotor speed, type of bearings and the duty of the machine. The mass of the rotor will, in the case of a machine with a horizontal rotor, play a dominant role in determining the load on the bearings and indeed, this will dictate the type of bearings to be used in a machine. The choice of bearing type is a crucial design decision and although it is anticipated that for the most part, readers of the present text will be operators rather than designers, for the interpretation of vibration characteristics, it is important to appreciate the behaviour of the main types of bearing.

1.3.1 Bearing Types

The most basic form of bearing is simply a brass bushing through which the shaft passes, involving a small air-filled clearance. This has the advantage of simplicity (and cheapness) and is suitable for small light machines. In its simplest form, the bearing relies on hydro-elastic effects from the rotation to generate lift forces, but more sophisticated versions employ some pressurisation of the air. Another variant of the air bearing incorporates a flexible foil within the clearance and this supports the rotor at low speed: as the rotor speed increases, hydro-elastic force dominates.

As the loading increases, it may become necessary to use a rolling-element bearing. These bearings are considerably more complicated but provide a robust support giving good stiffness properties, but only small damping terms. As the name implies, these bearings have a series of balls or rollers between the rotor and the stator, and hence, all contact between mating surfaces is essentially rolling rather than sliding. The minor component of sliding is accounted for in the lubricant. The rolling components are mutually located in a cage which itself rotates at a fraction (close to half) of shaft speed. However, this relatively complex construction means that a series of frequency components will be generated including cage frequency, ball spin frequency and, of course, the shaft frequency. The relative importance of these will vary with the condition of the bearing. Some very large machines, particularly low-speed examples such as wind turbines, are supported on rolling-element bearings.

At higher speeds, however, the majority of heavy machines are supported on oil journal bearings in which rotor and stator are separated by a film of oil. In hydrodynamic bearings (the over whelming majority), pressure in the oil film is generated from viscous stresses arising from the rotation of the shaft. Although the system is in essence much simpler that the rolling element bearing, the behaviour of the oil film is speed dependent and non-linear. In the plane bearing, sub-synchronous instabilities can arise but where appropriate, a modified type of bearing incorporating tilting pads can be used to overcome the problem. In subsequent chapters, there is considerable discussion of the phenomena arising with oil journal bearings. Their popularity arises from their high load capacity, but an additional desirable feature is the contribution to system damping. In large turbo-generators, the oil films account for about 99% of the total system damping.

Whilst these represent the most prominent forms of bearing, other types are gaining acceptance for particular type of application. The hydrostatic bearing is a type of oil-film bearing, but in this case, the pressure is supplied by an external pump rather than arising from the rotary motion. The provision of an external pump clearly adds complexity to the system, but this gives the significant advantage of a degree of control over the dynamic properties of the bearing. These units also have greater ability to operate at low speeds as compared to hydrodynamic units.

The ability to control the dynamic behaviour of a machine is currently an active research area and whilst hydrostatic bearings have a potential role here, much attention has been focused on active magnetic bearings (AMBs). In these units, the rotor is levitated in a magnetic field, the strength of which is controlled via a feedback loop using the measured shaft displacement as the controlling parameter. Hence, the rotor, bearing and feedback loop become a single entity and must be analysed as a system. AMBs have great potential but do have high cost. Because there is no contact and no fluid shearing (apart from air), friction is virtually zero and consequently, these bearings are suitable for very-high-speed applications. Damping can be introduced by adjusting the control algorithm. This is one of the great advantages of AMBs – the introduction of adaptive forces. The bearings are limited in load capacity, although there have been some applications in large compressors in gas pipelines. The fact that no lubricating oil supply is required presents a very significant advantage, particularly in remote locations. This is the motivation in recent development trends, aiming at completely electric aircraft and ships, but it is likely to be some time before these efforts reach fruition.

1.3.2 The Rotor

Turning now to the rotor, a marked trend towards flexible rotor is clear over the last half century. It is important to determine whether a rotor is essentially rigid or flexible as this can have a marked influence on machine behaviour and will prove crucial in assessing monitored vibration signals. Prior to World War II, all rotors were rigid and it was only with increased understanding that it was deemed feasible to operate machine with flexible rotors. Large alternators are now almost invariably flexible and an increasing number of smaller machines come into this category. Chapter 4 discusses the salient distinction between rigid and flexible rotors. In some respects, the differences are modest, but these distinctions can give rise to some marked changes in behaviour owing to the presence of higher modes and stress variations within the rotating elements.

1.4 Considerations for a Monitoring Scheme

Let us examine the logic which determines the type of monitoring system suitable for a given application. The prime question will be the capital value of the asset: if it is a low value asset which is readily replaceable, there may be little value in installing a complex, expensive system; rather, the requirement is mainly to detect an incipient failure and ensure a spare unit is readily available. The requirement is at stage one in terms of the classification scheme outlined earlier: purely statistical approaches can usually be applied to this problem, and this is often as simple as specifying that the vibration level must remain below a certain threshold level. A marginally more complicated system may consider a change in the shaft orbit, a problem which is discussed in Chapter 2.

The task of devising an appropriate monitoring scheme is complex and will depend on a number of operational parameters together with the range of anticipated fault conditions. Having decided to monitor this item of plant, an important issue is the frequency of monitoring – a very important issue which often carries major resource implications. To make an appropriate decision, the operator must consider the range of faults which occur on the type of plant in question and in particular the timescale on which the faults develop from inception to any threat to machine operation or integrity. The interval chosen for monitoring must be less than the development time of the faults or clearly, a damaging fault could be completely missed out by the system. On the other hand, monitoring too often generates excessive data which then required adequate systems and staff resources to make use of the data. This process of formulating the appropriate system has been formalised into a methodology called failure modes and effects analysis (FMEA).

A discussion of FMEA is outside the scope of this volume, but it has been widely discussed elsewhere.

For a slowly developing fault, or combination of faults, intermittent checking on a daily or even weekly basis may well suffice, and in such cases, it is likely that the required data can be logged by staff, either manually or increasingly with hand-held monitors which can then feed data centrally. It is unlikely that in such a case, the financial and other costs in implementing a dedicated computerised system could be warranted. The overriding consideration in the design of monitoring systems must be the cost-effectiveness, but the accurate evaluation of this is by no means a trivial task.

If particular faults on plant develop more rapidly, then some automation of the process becomes necessary or at least desirable. However, this involves a hidden cost: not only is there a cost involved in gathering the data, but also a not insignificant resource requirement in evaluating the data.

For high-value plant, the situation is rather different. In this case, there will rarely be an option of simply replacing the plant item and the monitoring system must address (at least partially) the issue of fault diagnosis with a view to optimising the repair strategy and minimising the so-called downtime. The requirement for a diagnostic capability almost inevitably implies a need for examining the frequency content of the signal and, hence, the need to record data at least several times per revolution of the shaft (actually, this is governed by Shannon's sampling theorem which is discussed in Chapter 2). Clearly such rapid sampling calls for a computerised system and with a profusion of data, appropriate strategies are needed to extract meaningful information which can make a real contribution to plant operation. Of course, there is still the decision to be made as to whether parameters are monitored continuously or at periodic intervals, but we need not dwell on this here.

Clearly, it is unrealistic to produce an accurate model of all machinery operations, although the later chapters do report significant progress in recent years towards this aim. Like many technical topics, one is forced into a regime of diminishing returns in which progress becomes limited by the restricted marginal benefit. It is appropriate to examine in general terms the realistic goals for any monitoring system. To make a meaningful contribution to plant operation, a monitoring system needs to provide full data in the event of an incipient fault, but this implies that overall, the volume of data produced is likely to be enormous and it becomes essential to devise some automated system to provide an initial filtering of the raw data. The direct automated diagnostics is somewhat beyond current capabilities and hence, the practical aim is for the system to alert the engineer (i.e. a human expert) of potential difficulties.

In preparing this text, efforts have been made to minimise reliance on mathematical analysis. However, as the guiding theme is to gain an in-depth understanding of machine behaviour, some mathematics is inevitable, but in all cases, this is basically expressions of Newton's Laws coupled with some degree of Fourier analysis. Whilst there is an element of non-linear behaviour in most systems, this is relatively minor in the majority of cases. Indeed, it is often the onset of this non-linearity which is the vital clue to establish a change. In all the discussions presented, the mathematical details are secondary to the theme: the central focus is the physical source of phenomena. The purpose of any theoretical or numerical model is to relate the observed behaviour to the processes within the plant item. Alternatively, this may be expressed as the conversion of data into information and it is worth considering this in a little more detail. If, for example, an operator has a vibration reading of a given amplitude and phase, this in itself is extremely difficult, if not impossible, to interpret. But there are a variety of ways in which this data can be rendered useful. Perhaps the most basic of these techniques is the formation of trends over time – and the period involved will depend on how a fault (or rather, a phenomenon) develops. The challenge is then to relate the change in behaviour to some change in condition. Clearly, this task may be made easier with complete records but only rarely will it be a trivial exercise.

The relationship between monitored behaviour and internal processes is often complex, and in this situation, it is necessary to utilise all information available. Clearly measured data contain information; this is not to say that the data are absolutely correct; indeed, it will inevitably incorporate noise and possibly some systematic errors, but it will contain some information. In the same way, a model of the machine, either analytical or numerical, will reflect some of the physics of the real machine and whilst containing some limitations, if used in the correct manner, it can be used to give additional insight. Chapter 7 discusses the incorporation of foundation effects using this type of 'blended' approach, but more generally, the resolution of machine problems often requires the use of information from a number of different sources.

The present state of knowledge in this field is far from complete, and in the next few years, there are prospects for significant advances. Over recent years, it has become possible to formulate reasonably accurate theoretical models to replicate the dynamic behaviour of practical machines. This has resulted partly from enhancement in our basic understanding and partly due to the advance in computing technology. Currently, the main features of behaviour are reflected in models, but with more recent identification techniques, it is becoming feasible to gain an insight into the more subtle distinctions between nominally identical machines. This is significant as this will make it possible to more fully utilise on-load data of machine operation. It has been a dilemma for many years that all plant operators have a vast quantity of data of their machines operating on-load, yet the information content of these records is relatively low because of a lack of interpretive capability. Only gross changes could be interpreted with any degree of confidence. A significant component of the variation was the inter-machine variation which is now better understood than in previous years.

1.5 Outline of the Text

In monitoring and diagnosing problems on rotating plant, a wide variety of parameters are assembled and gathered. Vibration data are commonly utilised, but because of their rapidly varying nature, the sheer volume of data leads to a question of presentation. This is not a trivial matter, because different formats of presentation bring out different features of the behaviour and it is frequently necessary to consider a number of different representations to gain a general overview of the phenomena. Chapter 2 discusses a wide range of the presentation of vibration phenomena and illustrates how various modes of presentation may be used to bring out different aspects of a machine's behaviour.

Chapter 3 discusses the fundamentals of machine modelling. After examining the need for theoretical models, the chapter outlines the FE method: it is fortunate that one of the simplest finite elements is also one of the most important in the representation of rotating machines. Having established the methodology, an explanation is given of the various ways in which data may be usefully presented. Often this is slightly different to the approaches presented in Chapter 2, because a different range of information is available. At the end of the chapter, we discuss the application of mode shapes and perturbation theory to enhance the understanding gained from models.

Chapters 4 and 5 give an extensive discussion of some of the more common faults on rotating machinery. It is not feasible to cover all possibilities in a single text, but the range covered here is representative. The rationale for this discussion is that it is simply not feasible to diagnose a fault without some idea of what the fault would look like, in terms of external parameters. Note that this is a distinguishing feature between detection and diagnosis. At what we might call level 1 of monitoring, the detection of a fault, it is quite possible, and often vital, to form a decision without any further knowledge. Such a simple system, usually based purely on statistical models or even just some threshold criterion, can trigger the removal of a machine from service or a reduction in load. However, as discussed earlier, to proceed to make a diagnosis, some insight into the physics of the situation and some form of model is needed. Chapter 4 discusses rotor imbalance, the single most prevalent fault in rotating machinery. The chapter begins with a detailed discussion of the differences between rigid and flexible rotors, as this has a number of important consequences and the methods used for correcting imbalance must be appropriate to the rotor class. After outlining the principal approach to rotor balance, discussion turns to rotor bends. As discussed in Chapter 4, it is important that bends and imbalance are not confused, even though they both give rise to synchronous vibration.

The survey of machine faults is continued in Chapter 5 with a discussion of misalignment, cracked rotors, torsion and rotor rubbing. These are complicated topics and some are still areas of active research. They are, however, important practical problems on real machines as well as being of substantial academic interest. The phenomena associated with misalignment remain incompletely understood even though it ranks second only to imbalance as a major problem on rotating machines.

Chapter 6 focuses on the influences of clearances in various types of rotating machines. There are many fine clearances occurring within machines usually filled with either working fluid or lubricant. The effect that these clearances have on the dynamics of the machine is typified by the case of a large centrifugal pump, and this is discussed at some length. The important issue is that the internal neck-rings act as subsidiary bearings and also have a profound effect on the internal pressure and flow distribution within the unit with consequent influence on the unit's overall performance. Clearances at blade tips in turbines, on the other hand, give rise to the so-called Alford's force. Actually, the behaviour of a shaft within an annular clearance is complex and a discussion of the range of phenomena and the ensuing machine faults are described. Later in the chapter, a discussion of the dynamics of recoil is given: this is another case in which clearances and wear patterns can play a profound part in determining both performance and dynamic characteristics of the system.

Chapter 7 is a little more specialised than the others and focuses on recent work on the problems of taking account of the dynamics of the supporting structure. On large machines in particular, such as turbo-alternators, the support can have a very significant influence on the dynamic properties of a machine and there are a number of factors which render conventional modelling of these structures virtually impossible, the reasons for which are set out in the chapter. These large machines present a special difficulty insofar as they are mounted on large complex structures whose own dynamic properties have substantial influence on the overall behaviour of the machine. In principle of course, all supporting structures will have some influence, the difference with these very large machines is that it is impractical to make their supports sufficiently stiff to minimise the effect. Because the supporting structures are complicated, it is impractical to obtain an adequate model of its influence on the machine simply by FE methods. Such an approach has been attempted by various authors over the last few decades, for example Lees and Simpson (1983) and Pons (1986). The overall conclusion is that although these models give general trends which can be useful by highlighting problems at the design stage, they are insufficiently accurate for use in monitoring and diagnosis of operational plant. This is because each support has a significant number of complex joints which are extremely difficult to measure even if complete data were available, but this is rarely if ever the case. The overall problem is illustrated by the fact that nominally identical units can have appreciably different dynamic conditions. Chapter 7 outlines recent progress in this area, showing how a dynamic model can be enhanced or updated in the light of measured data. These developments offer the prospect of significant progress in the next years as models may be sufficiently refined to study the variations in nominally steady-state conditions. Current methods only make very cursory use of this source of data.

Chapter 8 considers some of the more modern methods of treating data. These include ANNs and kernel density estimation. Both of these approaches provide a route to establish connection between data: they may be both regarded as non-linear fitting methods, but this can be invaluable in establishing relationships and thereby understanding. In this chapter, we also explore the methods for dealing with rapidly changing signals, all a part of the general field of time-frequency analyses. Some of these are still at the research stage, but practice over the next few years will do much to evaluate their usefulness. Kernel density estimation, wavelet transform, Huang–Hilbert transforms and expert systems are just some of the procedures discussed in this section, in addition to more model-based approaches. Emphasis is given on the potential of blending deterministic and more statistical-based approaches as a means of gaining the greatest insight into the operation of a piece of machinery. Indeed, this is the constant theme of the book – the constant effort to use available data to gain insight. In doing this, measured plant data and the predictions of theoretical or numerical models can contribute significantly to overall understanding.

Chapter 9 describes some case studies, taken from both the author's experience and the literature. The five examples given cover a range of machine faults and in most of the cases, a number of techniques are combined to arrive at a satisfactory solution to the problem. This is entirely realistic, as most practical problems demand some degree of ingenuity for their solution.

Finally, Chapter 10 is slightly more speculative than other parts of the text. An outline is given on a range of techniques currently available for both data analysis and model improvement. It is anticipated that a steady improvement in data analysis, perhaps in tandem with enhanced computing capabilities, will continue for some years and at the present time, it is not totally clear what the form of monitoring will be in the decades to come. Nevertheless, with due trepidation, some predictions are offered as to schemes which may become feasible (and cost-effective) in years to come. The culmination of developments in condition monitoring, diagnosis and control may combine to form what might be called a 'smart machine', that is a machine which will detect and diagnose an incipient fault, then apply a corrective force which will compensate for, or at least mitigate, the effects of the fault. There are several strands to research in this area and although it is still a little way from realisation, work is at a stage where progress may be rapid. A brief overview is given in Chapter 10 of the recent developments in this area.

Pending the fruition of these developments, work on the monitoring and diagnosis of problems in rotating machinery is a role that is vital across a wide range of key industries. It is hoped the techniques and fault descriptions presented here will aid that work.

1.6 Software

Throughout this text, the importance of a physical understanding of processes within a machine has been emphasised and almost invariably, this understanding requires some form of mathematical model. In some cases, these take an analytic form, but more usually, the complexity of real machines demands a software model. The work presented here uses the toolbox developed by Friswell et al. (2010), which is freely available at www. rotordynamics.info. This set of routines has been supplemented by a set of user interfaces and some other scripts for various common tasks are given. No other toolboxes are required and readers with access to MATLAB® 6 or later should have no difficulty in running any of the codes. Other readers may gain access by using the free SCILAB package. The real benefit of the numerical models is not specific answers to numerical case, but rather the ease of varying parameters and investigating the consequences of various scenarios. Readers are encouraged to experiment with the software as this offers a powerful learning experience.

References

Childs, D., 1993, *Turbomachinery Rotordynamics: Phenomena, Modeling and Analysis*, Wiley, New York.

Friswell, M.I., Penny, J.E.T., Garvey, S.D. and Lees, A.W., 2010, *Dynamics of Rotating Machines*, Cambridge University Press, New York.

Inman, D.J., 2008, *Engineering Vibration*, 3rd edn., Prentice-Hall, Upper Saddle River, NJ.

Lees, A.W. and Simpson, I.C., 1983, Dynamics of turbo-alternator foundations, in *Institution of Mechanical Engineers Conference*, London, U.K., Paper C6/83.

Pons, A., 1986, Experimental and numerical analysis on a large nuclear steam turbo-generator, in *IFToMM Conference 'Rotordynamics'*, Tokyo, Japan.

Price, E.D., Lees, A.W. and Friswell, M.I., 2005, Detection of severe sliding and pitting fatigue wear regimes through the use of broadband acoustic emission, *Proceedings of the Institution of Mechanical Engineers, Part J, Journal of Engineering Tribology*, 219(J2), 85–98.

Sikorska, J.Z. and Mba, D., 2008, Challenges and obstacles in the application of acoustic emission to process machinery, *Proceedings of the Institution of Mechanical Engineers, Part E, Journal of Process Engineering*, 222, 1–19.

2

Data Presentation

2.1 Introduction

The process of resolving problems on rotating plant, or simply monitoring performance, is in essence made up of two complimentary strands; first is obtaining plant data and second is interpreting it, either on the basis of experience or by the use of some form of theoretical model. In either case, the engineer is implicitly (or explicitly) fitting the measured data to a model or, to his/her understanding of the machine's operation, inferring the internal parameters of the machine being studied.

In this chapter, consideration is given to some of the basic forms of the measured data in the study of rotating machinery. In some respects, this is similar to the study of structural dynamics, but there is an important complicating factor; the variation of both the system and the forcing with shaft rotational speed. This is an obvious prerequisite to the material which follows in subsequent chapters.

We note that there are two fundamental distinctions between a structure and a rotating machine, as far as the dynamic behaviour is concerned. First, the dynamic properties of a rotating shaft depend on the rotational speed owing to gyroscopic terms and, in many cases, the property of the bearings. Second, the shaft rotation provides a potential energy source which implies that under certain conditions, vibrations can grow – a condition of instability. Both these issues will be addressed briefly in this chapter and further more extensively in Chapters 4 and 5. Prior to the detailed examination, however, these issues have some influence on the present discussion.

2.2 Presentation Formats

Although mathematical analysis of systems is discussed in Chapter 3, it is helpful to briefly mention the form of dynamic equations here. Any rotating machine can be described by an equation of the form

$$\mathbf{K}x + \mathbf{C}\dot{x} + \mathbf{G}\Omega\dot{x} + \mathbf{M}\ddot{x} = \mathbf{F}(t) \tag{2.1}$$

where Ω is the shaft speed and superscript '.' denotes time derivative.

The evaluation of the terms in this equation can be a little involved, but here, attention is focused on the structure of the equation. On the left-hand side, in addition to the stiffness, damping and mass terms (**K**, **C** and **M**, respectively), there is a gyroscopic term **G**. This is discussed in some detail in Chapter 3, but the essential point is that it arises from the conservation of angular momentum and an important consequence is that the dynamic behaviour of a rotor is dependent on the speed of rotation of the shaft.

In addition to gyroscopic effects discussed earlier, bearing properties often vary with shaft speed in a pronounced way. Because both the natural frequencies and mode shapes of a rotor vary with speed, the presentation of plant vibration data is somewhat more complicated than that for a non-rotating structure.

As mentioned in the introduction, there are two types of data measured on a typical rotating machine: (a) data from a steady running condition and (b) transient speed plots (run-ups and rundowns). On many types of plant, the former is plentiful and forms an important part of condition monitoring systems, but the transient data are extremely rich in information, but at a cost! The cost is in terms of difficulty of analysis.

2.2.1 Time and Frequency

The fundamental problem is that the dynamic properties of a rotating machine are functions of the rotational speed and indeed often functions of other parameters such as pressure or load. We consider first the dependence on speed, as this is (almost) invariably the key parameter. This variation with speed is of course one of the factors distinguishing a rotating machine from any general structures, and the variations may arise from the behaviour of bearings and seals as well as the effects of gyroscopic couples. The reader may note that gyroscopic terms tend to give rise to relatively small corrections for systems such as that shown in Figure 2.1 in which the rotor mass is between bearings. However, for situations where there is an appreciable overhang as is often the case with fans, as shown in Figure 2.2, gyroscopic terms may become very significant or even dominant.

At one time, vibration data would be recorded on magnetic tape for subsequent analysis. In essence, this was an analogue process, but the more modern

FIGURE 2.1
Basic rotor system.

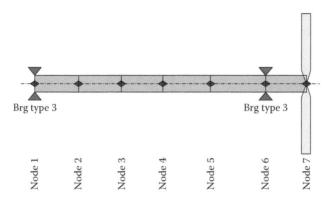

FIGURE 2.2
Layout of an overhung fan.

approach is to directly digitise data and store on a disc or other medium. It should be recognised, however, that any digitisation process involves decisions related to the form of the data. The sampling rate must be chosen to adequately capture the machine's behaviour. This choice will be governed by some assessment of the anticipated outcomes and considerations of aliasing and leakage problems (see, e.g. Bossley et al. 1999). There are two viable strategies for this; the first is to sample as rapidly as possible and then to resample in order to make evaluation viable. The alternative, and in some ways preferable, approach is to lock sampling to shaft rotation and sample at a number of times per shaft revolution. This strategy automatically eliminates leakage as a problem for most investigations. The essential point, however, is that important decisions have been made at this stage.

The stored data will be a time record of the vibration, and on occasions, this is in itself a key part of decisions on the machine's condition. More usually, however, the main focus of attention will be on the spectra content of the time records as this provides a route by which forcing and response frequencies can be related. In essence, this requires the use of the discrete Fourier transform (DFT), the digital equivalent of the Fourier transform. Given a set of measurements, x_r, evenly spaced in time, the frequency spectrum is given by X_k, and these are related by the DFT which is defined as

$$X_k = \sum_{r=0}^{n-1} x_r e^{j2\pi kr/n} \qquad k = 0,1,2,\ldots,n-1 \tag{2.2}$$

and the inverse is given by

$$x_r = \frac{1}{n} \sum_{k}^{n-1} X_k e^{j2\pi kr/n} \qquad r = 0,1,2,\ldots,n-1 \tag{2.3}$$

Note that these two relationships are analogous to the standard Fourier transform pair for continuous systems:

$$x(t) = \frac{1}{2\pi} \int_{-\infty}^{\infty} X(\omega)e^{j\omega t}d\omega \tag{2.4}$$

$$X(\omega) = \int_{-\infty}^{\infty} x(t)e^{-j\omega t}dt \tag{2.5}$$

The calculation of the DFT involves n^2 summations, and a much faster algorithm, the Fast Fourier transform (FFT), was reported by Cooley and Tukey (1965). This reduces the number of summations required to $(n/2)\log_2 n$. For large n, this represents a very significant reduction, and this algorithm has gained virtually universal acceptance.

The FFT algorithm, together with some subsequent refinements, has become a cornerstone of vibration analysis and signal processing in general. In the case of rotating machines, the shaft rotation speed is crucial, and the expression of data in the frequency domain clarifies the relationships of various components.

2.2.2 Waterfall Plots

In the analysis of structures, it is usual to present the spectrum of vibration by performing an FFT on the vibration signal, and this gives considerable insight into both the structure and the exciting forces. The same logic applies for rotating machinery, but there is an important distinction: in this case, both the forces and the structural properties are functions of shaft rotation speed and so a spectrum is required for each shaft speed. The difficulties in doing this are discussed a little later, but the result is a so-called waterfall plot, an example of which is shown in Figure 2.3.

The scale labelled 'number' can be rotor speed or time or indeed other parameter such as load. The waterfall plot, although a fairly crude device, offers a comprehensive overview of the system dynamics, and in many instances, it will form a prelude to more detailed study. It is interesting to make some observations on this particular plot. The data were taken from a large laboratory rig at Swansea University which is shown in Figure 2.4. From this simple plot, two features are immediately obvious. In addition to the main diagonal line showing synchronous excitation arising from imbalance, there are a number of other lines appearing corresponding to excitation at harmonics of shaft speed. These undesirable features must arise from non-linearity in the system, and this was quickly ascribed to the flexible coupling between the motor and the shaft. At the time, this coupling was of the single-membrane type which proved inadequate for the purpose. On the evidence

FIGURE 2.3
Sample waterfall plot. (Reproduced with permission from the Institution of Mechanical Engineers. Published as part of Lees, A.W. et al., *Mech. Syst. Signal Process.*, 23(6), 1884, 2009.)

FIGURE 2.4
Experimental rig – Swansea University. (Reproduced with permission from the Institution of Mechanical Engineers. Published as part of Lees, A.W. et al., *Mech. Syst. Signal Process.*, 23(6), 1884, 2009.)

of this plot, the coupling was replaced with a double-membrane type, and the harmonic excitation was dramatically reduced. There will be unavoidable sources of non-linearity in many machines, but an understanding of their effect is a prelude to an adequate correlation with any model.

The second feature of interest on this waterfall plot is the (nearly) vertical line of peaks corresponding to about 300 Hz. This decreases with shaft speed and is therefore likely to be associated with a backward whirl mode of the rotor. Somewhat less clear, there is evidence of the corresponding forward mode slowly increasing with speed from the same value around 300 Hz. This excitation is not locked to shaft speed and arises from a resonance in the supporting structure. On real power plant, this represents a major influence on machine vibration properties. Modelling the foundation is in principle straightforward; the (major) difficulty is in having the appropriate data to use as each individual foundation has its unique properties. Much lower down in frequency, there is a resonance at 48 Hz which corresponds to a critical speed of the system and which is largely independent of speed.

The details are not particularly important, but the key point here is that so much information may be gleaned from a waterfall plot of a single channel. Added information may be presented through similar plots from other channels of data. Recent work on this problem is discussed in Chapter 7. The interested reader may also refer to a recent review (Lees et al. 2009).

2.2.3 Scatter Plots (or Carpet Plots)

Just as the waterfall plot provides a simple 'broad-brush' overview of behaviour at a range of speeds, the so-called scatter plot is a useful tool for viewing on-load data at constant speed. An important part of vibration monitoring concerns the analysis of data whilst the machine is operating, very often at a nominally constant speed. Whilst this is a regime in which there is inherently less information available from the data, since it is the state of normal operation, there is a vast amount of data available at essentially no cost. However, the vast data resource contains limited information, and it is important to extract the limited information which is accessible.

It is informative to commence our discussion by posing the question 'Why does the vibration behaviour vary over time when the rotational speed is fixed'? The precise answer to this will depend on the type of machine being studied, but for the sake of illustration, we will discuss the case of a large turbine generator. The variation of the magnitude of vibration on the front bearing of the LP turbine is shown in Figure 2.5, but this only tells part of the story. Nevertheless, it can be seen that the variations are not insignificant.

This figure shows the amplitude of the vibration velocity at 2 min intervals over a period (about 11 days). For the vast majority of this period, the machine

FIGURE 2.5
Bearing vibration over an extended period.

was operating at a nominally constant speed of 3000 rpm. Despite this 'steady operation', the vibration amplitude shows considerable variations, and this raises the following two issues:

a. Can any insight be gained from theses variations over time?
b. Can the data be sensibly used for condition monitoring purposes?

The naïve answer to the second question is clearly 'yes' since if the vibration undergoes a large change, this would clearly invoke some investigation as to the cause so we repose this question as 'How small a change in vibration levels can be considered as significant'?

First, however, we address the more fundamental question of why these variations arise. The answer is rather complex, and the details are still the subject of active research work.

Some of the variations which influence the machine are

a. Ambient temperature
b. Water level in the condenser
c. Steam temperature
d. Steam pressure
e. Variations in power supplied

f. Variations in reactive power

g. Rotor and stator cooling operation

h. Rotor current settings

and many more: for each of the data points shown, 54 psuedo-static parameters are recorded. Of course, not all of these are equally important. A method for deciding the importance of different parameters is introduced in the succeeding text. For instance, at coastal power stations, the state of the tide has been observed to have some influence over the dynamics, no doubt by effecting changes in the water table. For other types of machine, the list of factors will be different, but there will invariably be features which give rise to variations during nominally constant operation. A different perspective is gained if we re-plot these data showing the real part against the imaginary part. The result is Figure 2.6. This is a scatter or carpet plot. The measured vibration velocity at each of the 2 min intervals is plotted on the Argand diagram, thereby representing both magnitude and phase. Note that ideally, under constant running conditions, all the points should coalesce, but because of fluctuations, there is also some spread. The problem is to assess whether this spread is within normal bounds.

This can be simply a plot of the time data of the motion in two orthogonal directions. The real and imaginary parts of the synchronous term could also be plotted in this way, but this complication is unnecessary. The simple plot gives an overview of long-term variations.

FIGURE 2.6
Typical carpet plot.

The precise mechanism of the various interactions is rather beyond the current state of machine modelling, although it is hoped that recent developments will lead to progress in this area. For the present, however, statistical approaches are used to gain some insight into on-load data. This plot shows some interesting structure, but the general properties are first discussed using a simulated example.

One advantage of using a scatter plot is that, despite its simplicity, it gives information on the complete vibration signal, in two directions. Interpretation is, however, slightly involved. Consider the data shown in Figure 2.6: these are data from a real turbine, and the overwhelming majority of the data points are clustered in a small area of the diagram. Two questions arise, (a) which, if any, of the outlying points represent a plant excursion of significance and (b) what is an acceptable level of spread in the distribution. A statistical approach is followed here.

Figure 2.7 is simply simulated data using Gaussian random numbers which we use to illustrate some manipulation methods. Not surprisingly, no structure is evident, and it is straightforward to generate circles which denote 1 and 3 standard deviations from the average. These circles, superimposed on the data, can be used to assess the significance of any deviant behaviour in purely statistical terms. But real data, having structure, are somewhat more complicated to deal with: a simulated case is shown in Figure 2.8. In this case, it is clear that the circles drawn to represent 1 and 3 standard deviations bear little relationship to the data being studied. In using these values as an assessment tool, predictions would be totally

FIGURE 2.7
Scatter plot of 2D Gaussian noise.

FIGURE 2.8
Simulated correlated system.

inconsistent, sometimes being unduly restrictive, whilst others being too lax. The problem now is to determine a way in which the collection of data can be adequately represented. In Figure 2.8, the use of circles to define the spread of the data is clearly inappropriate, not only do the two directions scale differently, but also there is an issue with the orientation of the bulk of the data.

To resolve these issues, consider the set of data points x_i, y_i. The first step is to take averages $\bar{x}, \bar{y}$ and use these to calculate the variance matrix **A** as

$$\mathbf{A} = \frac{1}{N} \begin{bmatrix} \sum_{i=1}^{i=N} (x_i - \bar{x})(x_i - \bar{x}) & \sum_{i=1}^{i=N} (y_i - \bar{y})(x_i - \bar{x}) \\ \sum_{i=1}^{i=N} (x_i - \bar{x})(y_i - \bar{y}) & \sum_{i=1}^{i=N} (y_i - \bar{y})(y_i - \bar{y}) \end{bmatrix} \quad (2.6)$$

This matrix is now used to resolve the data into principal components, essentially transforming the data into a new co-ordinate system. The key step in this is the solution of the eigenvalue problem

$$A\Phi = \Lambda\Phi \quad (2.7)$$

where Λ is a diagonal matrix of eigenvalues and Φ is the matrix whose columns are the eigenvectors. Note that the two eigenvalues of the correlation matrix give the two deviations (the major and minor semi-axes of

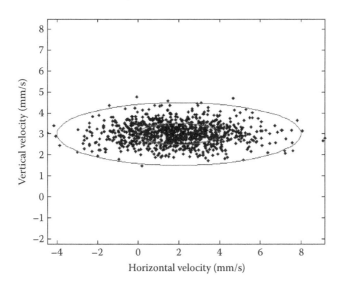

FIGURE 2.9
Plot of orthogonalised data.

the bounding ellipse), whilst two mode shapes determine its orientation. The data are now modified by the transformation

$$\left\{ \begin{matrix} X_i \\ Y_i \end{matrix} \right\} = [\Phi] \left\{ \begin{matrix} x_i - \bar{x} \\ y_i - \bar{y} \end{matrix} \right\} \tag{2.8}$$

The transformed data are plotted in Figure 2.9, and it is clear that the dependence of each data point on the new co-ordinates (X_i, Y_i) is now independent. This being so, it is straightforward to calculate the variances in the two orthogonal directions as (since clearly, $\bar{X} = \bar{Y} = 0$)

$$\sigma_x^2 = \sum_{i=1}^{i=N} \frac{X_i^2}{N} \quad \sigma_y^2 = \sum_{i=1}^{i=N} \frac{Y_i^2}{N} \tag{2.9}$$

and using this information, ellipses may be drawn around the data showing the extent of variability at one and three standard deviations. Note that the two variances shown earlier are just the two eigenvalues of the matrix **A**. Transforming everything back to the original co-ordinates, the result is Figure 2.10. The important point to note here is that the ellipses representing deviations from the norm are now far more representative of the data set. Hence, any judgment as to whether or not a change has taken place is far more accurate using these curves. The circles of Figure 2.8 would yield totally inconsistent warnings.

FIGURE 2.10
Analysed scatter plot.

The prediction and understanding of the subtle variations revealed by these plots are beyond the scope of current modelling techniques, although the recent developments discussed in Chapter 7 offer some prospect of advances in this area. A significant difficulty has been that, in practice, no two machines are identical and the changes due to damage are often commensurate with the differences between nominally identical machines. One possible way forward in the understanding of 'steady load' conditions is the use of artificial neural networks and other techniques which are somewhat loosely termed artificial intelligence. In some ways, this might be considered to be somewhat against the theme of this book where physically based modelling is advocated in the quest to understand dynamic behaviour. However, physics-based and statistically based approaches are not necessarily opposing methods.

Further information can be gleaned from the study of the scatter plots; the methods used are discussed in Chapter 8.

2.2.4 Order Tracking in Transient Operation

In most, if not all, turbo-machinery, the predominant frequency of excitation is related to the rotational speed of the shaft. This implies that during transient operation, that is during run-up to operating speed or coast down to rest, a broad range of frequencies is 'excited', and hence a comprehensive assessment of the machine's dynamic properties is available from a careful treatment of the monitored data. Much of the power of run-down curves derives from the fact that they can easily relate to predictions from theoretical (almost always FE) models: it is often the comparison of models and plant data which really contributes to a thorough understanding of plant behaviour.

Whereas the scatter plot gives overall trends over a period, the run-down curve gives time-specific information of frequency resolved information. The plots shown in Figure 2.11 show the vibration occurring at shaft speed, often referred to as 1X (and this, of course, is the component of vibration which may be excited by rotor imbalance). Analogous plots may be produced by 2X and higher orders. However, it becomes apparent that for a large turbogenerator with 14 bearings, each vibrating in two directions, a very considerable number of plots are generated given that one always requires first and second orders but often 3X and 4X in addition. The process of resolving vibrations into shaft speed-related components is known as order tracking, and this plays a crucial role in machine diagnostics.

This pair of plots (Figure 2.11) depicts the magnitude and phase of the response as a function of shaft rotation speed during transient operation (either run-up or rundown) resolved into shaft orders, with the implicit assumption that this reflects the quasi-static situation. The curves give considerable insight into the condition of the machine at the time of the transient,

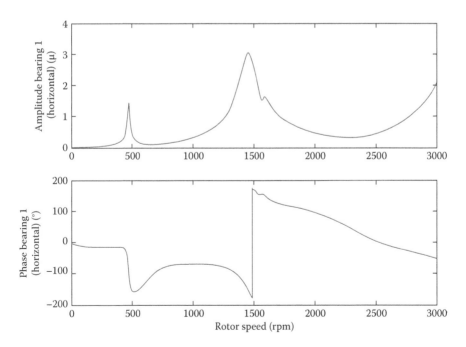

FIGURE 2.11
Typical rundown plot.

and this is extremely useful in machine diagnostics. Technical difficulties in obtaining reliable transient plots arise from the conflicting requirements of the sampling time needed for good frequency resolution and the localisation of individual shaft speed.

Both amplitude and phase show significant variation with shaft speed, and Figure 2.11 shows the synchronous components. Similar plots can be made for the higher orders, the twice per revolution being particularly useful. The problem arises of how to calculate the frequency response since a straightforward Fourier transform assumes that the system is time invariant. The simplest, and still the most commonly used, technique is the short-time Fourier transform, but two difficulties arise as we now illustrate. The first difficulty is the problem of treating the data in a meaningful way to give an accurate reflection in the frequency domain. The problems arising in transient studies is discussed by Fyfe and Munck (1997). The second issue is to relate this transient behaviour to a quasi-static situation by reflecting on how rapidly the machine can respond.

2.2.4.1 Acquiring the Transient

Figure 2.12 shows part of the FFT of a sinusoidal signal at 25 Hz over 2 s. Notice there is a single line in the FFT at 25 Hz. Also shown on the figure is the same signal but taken over 2.02 s, and the FFT looks rather different: although the sampling time has been longer, an inferior result has been obtained because of the phenomena known as leakage. This occurs because in taking an FFT of a signal over a finite time interval, the signal is

FIGURE 2.12
A comparison of the spectra over 2 and 2.02 s.

implicitly repeated throughout all time to create a signal of infinite length and this gives rise to inaccuracies if the beginning and end of the measure signal do not join continuously. Another way of saying this is that the measured sample must contain an integer number of cycles. In rotating machinery, the way around this problem that is normally used is to fix the data acquisition with relation to shaft position rather than time. This is highly relevant in that in most cases the principal concern is to relate the vibration signal to rotation speed rather than an absolute frequency reference. This locking can be achieved either in the data acquisition hardware or in software by digitally re-sampling. Before proceeding, we note some of the salient features of Figure 2.12. The resolution is, of course, set by the overall sample period, and since this changes by only 1%, no difference in resolution is apparent. The major difference in the second case is that the time span does not represent an integral number of cycles. The graph of this case shows that the peak is now close to (but not equal to) 25 Hz, and its magnitude has decreased to a considerable degree. On the other hand, the value of the FFT at other frequencies has increased. The distribution has broadened: the 'signal' which should all be at 25 Hz has 'leaked' into neighbouring frequency bands.

If a structure is being studied, analysis is quite straightforward – one simply samples the motion over a period chosen to give the required frequency resolution from

$$\Delta f = \frac{1}{T} \qquad (2.10)$$

with frequency expressed in Hertz and T is the period over which the data are sampled. Equation 2.10 has a simple physical meaning; the maximum resolution which can be observed corresponds to a difference of a single cycle over the period monitored.

With machinery in transient operation, the situation is more difficult in that, during the interval T, the speed of the machine, and consequently the excitation, will change. Hence, the choice of the sampling period will sometimes be a compromise. Let us assume that a machine runs down to rest exponentially so that the speed, Ω, is given by

$$\Omega = \Omega_0 e^{-\alpha t} \qquad (2.11)$$

This expression is only approximately true, but it will suffice for the current discussion. Therefore, we may write

$$\frac{d\Omega}{dt} = -\alpha \Omega \qquad (2.12)$$

In this context, it is rather more meaningful to express the sampling interval in terms of the numbers of shaft rotations, N, and then the decrease in speed during this interval is given by

$$\Delta\Omega \approx -\alpha\Omega T = -\alpha\Omega\frac{2\pi N}{\Omega} = -2\pi\alpha N \tag{2.13}$$

The frequency resolution (in Hertz) is given by

$$\Delta f = \frac{1}{T} = \frac{\Omega}{2\pi N} \tag{2.14}$$

Hence, Equations 2.13 and 2.14 yield two conditions which tend to conflict: Equation 2.13 expresses the fact that given a long sampling time will allow the shaft to decelerate, hence degrading the frequency resolution, whilst on the other hand, Equation 2.14 shows the need for a long sample period to maximise the frequency resolution (i.e. to minimise Δf).

Putting these two conditions together, the best resolution possible is achieved when

$$\Delta f_1 = \frac{\Delta\Omega}{2\pi} = -\alpha N = \Delta f_2 = \frac{\Omega}{2\pi N} \tag{2.15}$$

$$\therefore N = \sqrt{\frac{\Omega}{2\pi\alpha}} \tag{2.16}$$

Using this value, we see that

$$\Delta f = \sqrt{\frac{\alpha\Omega}{2\pi}} \tag{2.17}$$

Therefore, as the speed decreases, the frequency resolution improves: the vibrational properties at very low rotor speeds are of less significance as any forces arising are low. Equation 2.16 also shows that for high values of α, the slew rate, frequency resolution is poor, and this makes conventional run-down methods inaccurate. For applications involving high rates of change of rotational speed, a rather different method has been developed which involves the use of Kalman filters. This approach is somewhat more involved and is discussed in Chapter 8.

To illustrate the method in Equations 2.10 through 2.16, consider a gas turbine used for power generation: the graph of rotor speed as a function of time is shown in Figure 2.13.

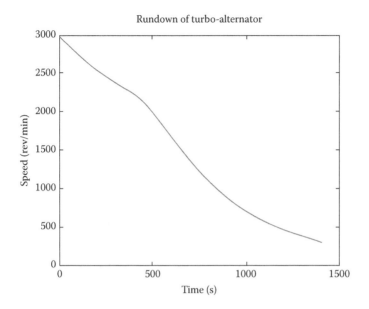

FIGURE 2.13
Speed decay curve.

Clearly, this curve is not exponential, but it is not too far away and an estimate for the effective α value is readily obtained by considering this curve, in
particular the initial deceleration rate, in this case about 6 rev/min/s.

To effectively assess the transient behaviour, the operator has a dilemma:
in order to maximise the frequency resolution, a long time block is
required, but an excessive period will bring errors due to the machine's
speed variation. Therefore, we seek an optimum compromise and, rather
than consider time explicitly, consider the block in terms of a number of
complete shaft rotations (this automatically resolves any possible leakage
problems).

If we sample over N cycles, with a time frame T, the resolution is given by

$$\text{res} = \frac{1}{T} = \frac{\Omega}{2\pi N}$$

But in this time, the shaft slows by

$$\Delta\Omega = -\alpha\Omega T = 2\alpha\pi N \quad (\text{in rad/s})$$

$$\frac{\Delta\Omega}{2\pi} = \Delta f$$

Hence, from Equation 2.15

$$\frac{\Omega}{2\pi N} = \alpha N$$

The value of α is effectively given in the question: α is just the initial decelera-tion divided by the initial speed (in consistent units), that is $\alpha = \dfrac{6}{3000} = \dfrac{1}{500}$. Using this yields

$$\therefore N = \sqrt{\frac{\Omega}{2\pi\alpha}} = \sqrt{\frac{50}{1/500}} = 50\sqrt{10}$$

$$\Rightarrow N \approx 158$$

$$= 158 \text{ rotations of the shaft}$$

Since we typically need four harmonics, eight samples per rotation are needed, that is $8 \times 158 = 1264$ samples in each window of data. The resulting frequency resolution is given by

$$\Delta f = \frac{\Omega}{2\pi N} = \frac{100\pi}{2\pi \times 158} = \frac{50}{158} \approx \frac{1}{3} \text{ Hz}$$

It is very straightforward to check the numbers here: sampling over 158 revolutions will take 3.16 s during which time the speed reduces by about $6 \times 3.16 = 19$ rev/min or $19/60 \approx 1/3$ Hz; hence, the two conflicting effects yield the optimum compromise.

Table 2.1 shows the best resolution that can be attained using this approach for a range of rundown rates. The initial speed is taken to be 3000 rev/min. The time quoted is that taken to run down to 10% speed.

Although in many cases an exponential speed decay curve will be appro-priate, it may be observed that Figure 2.13 deviates significantly from this ideal, and it would be equally appropriate to idealise this curve as linear. Exactly the same logical argument may be applied to estimate the appropriate

TABLE 2.1

Rundown Frequency Resolution

Time (s)	α	N (revs)	Δf (Hz)
10	0.23	15	3.4
50	0.046	33	1.5
100	0.023	47	1.1
500	0.0046	104	0.48
1000	0.0023	147	0.34

rundown parameters. Denoting the speed S in rev/s (as it makes things a bit clearer than rev/min), then $S = \Omega/2\pi$ and

$$\frac{dS}{dt} = -\frac{2}{60}$$

The time interval for N cycles is (initially) $N/50$ s, and in this time, the shaft speed will reduce by $\Delta S = -\frac{2}{60} \times \frac{N}{50}$. Consequently, the optimum sampling occurs when

$$\Delta f = \frac{50}{N} = \frac{2}{60} \times \frac{N}{50}$$

or

$$N = 50\sqrt{30}$$

Note that this figure is not substantially different from the value used earlier, and it may be inferred that the precise detail of the rundown rate is not a very sensitive determinant of the optimum sampling.

As mentioned above, this approach is inappropriate for use on machines which change speed rapidly. The alternative method, the Vold-Kalman approach, is named after its inventor Harvard Vold who developed the technique using Kalman filters.

2.2.4.2 System Response

At this stage, we turn our attention to another aspect of the interpretation of transient vibrational signals: clearly, this has to be considered before any sound judgements can be made on the condition of a piece of machinery. To gain some insight into the rate at which a machine can reach the response levels of steady running, we consider two cases: the first is the response to a step (constant) force, whilst the second case is the initiation of a sinusoidal excitation.

The transient response of a single degree of freedom system to a unit step change is readily calculated (using Laplace transforms) as

$$c(t) = 1 - \frac{1}{\beta}e^{\zeta\omega_n t}\cos(\omega_n\beta t + \theta) \tag{2.18}$$

where $\beta = \sqrt{1-\zeta^2}$ and $\theta = \tan^{-1}(\beta/\zeta)$. ω_n is the (undamped) natural frequency, and ζ is the damping ratio (i.e. the fraction of critical damping).

Figure 2.14 illustrates the response to a unit force for various values of damping. Notice that there is a characteristic time of the response $\tau = 1/\zeta\omega_n$.

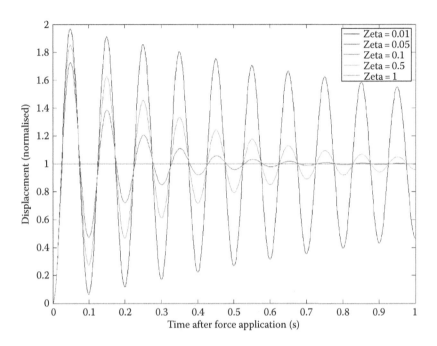

FIGURE 2.14
Response to a step force.

The same factor occurs if we now consider the effect of an oscillator starting an oscillator from rest. Clearly, it will take time for the response to develop to the steady-state response. The equation of motion can be expressed as

$$m\ddot{y} + c\dot{y} + ky = me\omega^2 \cos\omega t \tag{2.19}$$

where e is the mass eccentricity. This may be re-written in the usual form as

$$\ddot{y} + 2\zeta\omega_n\dot{y} + \omega_n^2 y = e\omega^2 \cos\omega t \tag{2.20}$$

The solution is readily written in the form

$$y = \frac{me\omega^2 \cos\omega t}{\sqrt{\left(\omega_n^2 - \omega^2\right)^2 + 4\zeta^2\omega^2\omega_n^2}} + e^{-\zeta\omega_n t}\left(A\sin\omega_d t + B\cos\omega_d t\right) \tag{2.21}$$

where the damped natural frequency is given by

$$\omega_d = \omega_n\sqrt{1 - \zeta^2}$$

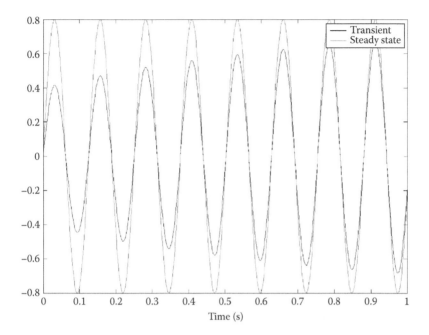

FIGURE 2.15
Convergence to steady state with $\omega_n = 50$, $\xi = 0.025$.

In practice, the difference between the damped and un-damped natural frequencies will rarely be of concern. It is now a straightforward, if tedious, task to determine the constants A, B to give the complete solution. Clearly, the approach to the steady-state solution is governed by the time scale $1/\zeta\omega_n$. An example of the convergence rate is shown in Figure 2.15.

If we consider now the time required for the response to be within 1% of the steady-state value, then

$$\zeta\omega_n t = 2\pi\zeta m = \log_e(100)$$

where m is the number of complete cycles. For a typical example, consider 2% damping in a system with $\omega_n = 80\pi$. This gives a value for m of 37 cycles which at 50 cycles per second represents 0.75 s for the system to saturate. Provided the machine in question does not reduce speed significantly over this period of time, quasi-static analysis is appropriate. It is worth noting that an un-damped system will never settle to the steady-state value. To analyse the general system in the general case, resort must be made to consider the general response in the time domain.

At this stage, it will be clear that for the analysis of rapid transients, there are two quite distinct difficulties. First, there is the technical problem of obtaining an adequate time-frequency description, and this is discussed in

Section 2.2.4.1, but the quite separate issue is how the system responds to transient forcing.

The general response is described by

$$h(t-\tau) = \frac{e^{-\zeta\omega_n(t-\tau)}}{m\omega_d}\sin\omega_d(t-\tau) \quad \text{for } t > \tau \tag{2.22}$$

The displacement of the rotor may be written as

$$y(t) = \int_0^t h(t-\tau)\left[e\omega^2\cos\omega\tau + e\frac{d\omega}{dt}\sin\omega\tau\right]d\tau \tag{2.23}$$

Using the convolution theorem, the Fourier transform of this equation can be readily formed (although this may involve substantial computing – the precise requirements will be examined a little later). Knowing the Fourier transforms Y and F, the function H is readily calculated. This implies that we have overcome the difficulties of what may be termed 'unsaturated response', and this is indeed the case because full use has been made of the machine's dynamic response via the impulse response function. In practice, most analysts assume an approximation to saturated conditions as this sim-plifies computation considerably. Given a validated model, a simulation of a defined transient can be simulated although the marginal improvement in accuracy means this is rarely justified.

2.2.5 Shaft Orbits

The effective evaluation of machinery inevitably involves a wide variety of techniques, both sophisticated and reasonably simple. An important maxim of the analyst must be 'never neglect the obvious'. An extremely simple mea-sure of plant performance is a plot of an orbit of either the shaft or bearing motion (as both can offer valuable insights). This is simply a matter of plot-ting x against y over one or preferably a series of shaft cycles.

Orbits can take a number of forms, and we do not attempt a comprehensive discussion here. Figure 2.16 gives four possible orbits, and these can help give considerable insight into the operation of a machine. Like most techniques in machine diagnosis, the shape of the orbit will rarely give conclusive evi-dence of some machine fault, but it will very often provide an important part of a collection of interlocking pieces of evidence. So what can be inferred from the orbits shown in Figure 2.16? An important point to note is that all four figures are records at a fixed shaft rotation speed. Figure 2.16a shows a purely circular orbit, and this may suggest that the machine is symmetric, in the sense that it is equally stiff in the two directions orthogonal to the axis of the shaft. To be conclusive about this, one would need to check the orbit at

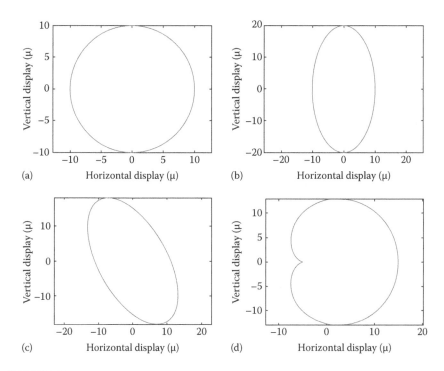

FIGURE 2.16
Some common orbit shapes: (a) orbit of symmetric machine, (b) orbit of asymmetric machine, (c) orbit of coupled machine and (d) orbit of non-linear machine.

some other shaft speed; a symmetric machine will show circular orbit at all speeds. The orbit also reveals that the forcing is harmonic and not predominantly in one direction.

In Figure 2.16b, the orbit has changed to an elliptical form, and this shows clearly that the machine is not equally stiff in the two orthogonal directions. It would be tempting to suggest that the machine is more flexible in the vertical direction, but more evidence is needed. At first sight, this may be a little surprising, but given that the stiffness is asymmetric, natural frequencies are different for vertical and horizontal directions. It is possible that the speed chosen happens to be close to the vertical resonance. Information at other speeds would be needed to reach conclusions on the relative stiffness in the two directions.

Figure 2.16c again shows an elliptic orbit, but the ellipse is tilted. This suggests some cross-coupling between the x and y directions. This could arise in the supporting structure, the bearings (particularly oil film journal type) or gyroscopic effects, but the latter tends to be significant only in overhung systems. Finally the last orbit, Figure 2.16d shows a marked change in that an internal loop has appeared. This implies the system has some non-linearity or, less often, some external forcing at a frequency other than that corresponding to the operating speed. Common sources of non-linearity are oil bearings, shaft rubs, gearboxes and (less often, fortunately) shaft cracking. These features will be discussed in greater depth in Chapter 5.

2.2.6 Polar Plots

Plant engineers frequently use the polar plot to assess the performance of a machine over a range of rotational speeds. Rather than plotting data in terms of the real and imaginary parts as in the scatter plots, plotting is in terms of magnitude and phase although, of course, the two representations are entirely equivalent. Recordings are made at a series of rotational speeds, but the plot is not extended over a long period of time as was the case with scatter diagrams. The simple reason for this is the attempt to encapsulate different aspects of the dynamic properties. Figure 2.17 shows an example of a polar plot.

The points on the curve shown represent the magnitude and phase of vibration (at a given measurement point on the machine) for different shaft speeds. Clearly, this implies that we are referring to synchronous vibration components. From the curve shown, we immediately see the variation of response with frequency throughout a transient operation (either run-up or rundown). Each point on the curve may be associated with

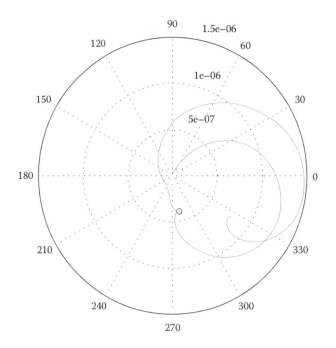

FIGURE 2.17
Typical polar plot.

an operational speed. Any critical speed close to running speed would normally give rise to substantial phase changes.

The information on the polar plot is precisely the same as that on first-order rundown plot (transient response): it is just presented in a compact form. This is a perennial problem – the presentation of the vast amount of data which emerges from rotating machinery. This plot is taken from a gas turbine undergoing a rapid run-up to speed. The three circular shapes (i.e. two large diameter circles and a single small one) are indicative of resonances.

2.2.7 Spectragrams

Another case of using a single diagram to give an overview is the spectragram, and an example is shown in Figure 2.18. This particular example shows some acoustic emission hence the high frequencies shown, but exactly the same technique may be applied with good effect to vibration data. This plot is closely related to the waterfall plot, with a series of FFT plots arrange for each time frame. The heights at each point are normally denoted by colour, but in this instance, it is shown in shades of grey. Clearly, the same time resolution problems, as discussed elsewhere, must be taken into consideration, in preparing such a plot. This is a case of time-frequency analysis, and the wider problem is discussed in Chapter 8. It is another example of the presentation of a significant volume of information on a single plot. Whilst for many purposes there may be a need to focus on more specific details, it can be extremely helpful in defining the context of an investigation.

2.3 Comparison with Calculations

Section 2.2 discusses some of the more common ways of presenting data, and an outline is given there of some of the physical principles involved in the interpretation of the measured signal. The important step is of course the use of these data to infer the condition of the machine by detecting the physical processes which give rise to the measured parameters. Expressed another way, the engineer is inferring information on internal parameter from external measurements: furthermore, the measurements will be subject to some errors and noise. This step is far from simple and is sometimes not accorded the attention it deserves.

On rare occasions, one may be able to measure precisely a parameter of interest, but more usually, information is inferred from the generality of some other data. This process of interpreting data relies on a variety of sources, including knowledge of similar machines, records of previous operation, maintenance records and mathematical models, and in the last few decades, the importance of mathematical models in this interpretation has increased

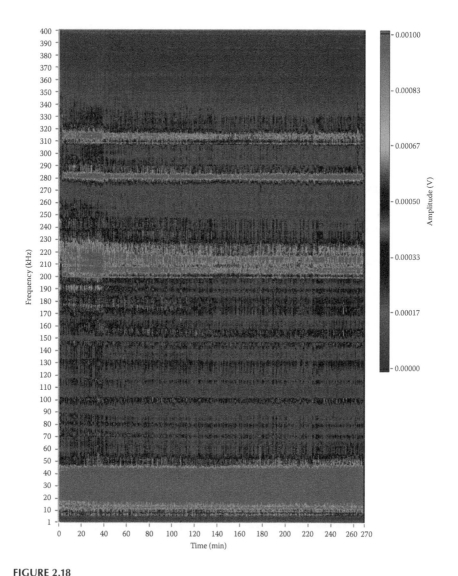

FIGURE 2.18
Spectra-gram showing acoustic emission data. (Reproduced with the permission of the author from Quiney, Z.A., Use of acoustic emission for bearing monitoring, Ph.D. thesis, Swansea University, Swansea, U.K., 2011.)

dramatically. Of course, this is no substitute for records that are available, but the key benefit of an adequate model is the ability to probe questions such as 'what if'? The word 'adequate' applied to a model is an interesting concept which is explored elsewhere in this text, particularly in Chapter 7, in which measured data are used to enhance the model.

It is worth emphasizing that the model, of whatever form, should be regarded as a tool to increase one's overall understanding of the machine in question. Since it incorporates our (albeit limited) physical understanding of the operation, it should be used to add to the overall information available. Just as measurements are subject to some uncertainty, so is the model of whatever form.

Many of these models will be based on the finite element method which is discussed in detail as applied to rotor dynamics by Friswell et al. (2010). A more general discussion of the approach is given by Irons and Ahmad (1980) among many other books on the topic. The authors do, however, give a very appropriate overview in the preface 'Computing is for understanding, not numbers'. This is a sentiment with which the present author is in total agreement.

2.4 Detection and Diagnosis Process

In monitoring and diagnosis of plant problems, there is considerable variety, and each instance will have particular features. Although the discipline cannot be reduced to a routine or checklist, some common theme do emerge as useful patterns. Comprehensive records are, of course, absolutely invaluable, and with modern data storage technologies, this is easily implemented. Section 2.2.2 outlines some of the simple statistical approaches that may be applied to long-term data, and these methodologies may give the first indication of some plant 'excursion' that warrants investigation. Experience may well dictate machine shutdown under certain conditions. In general, however, the statistical approach will just highlight an area or event to be examined.

To take matters further usually (but not always) requires a study of the data in the frequency domain enabling the engineer to relate phenomena to the operation of the machine. The most common faults occurring in turbomachines are discussed in Chapters 4 and 5, but here, some rough outlines are offered. If excitation is at synchronous shaft speed, the most common problem is imbalance, although (less commonly) it may also indicate a bent rotor. These two effects can only be distinguished by further examination as outlined in Chapter 4. Higher-order excitation indicates some form of non-linearity, possibly a rotor crack or misalignment. In either case, further investigation will be needed probably involving the analysis of machine transients.

The presence of components of vibration which are not multiples of shaft speed normally indicates some non-linear behaviour of the system. If it is below half rotor speed, in machines with oil bearings, then oil whirl may be suspected, and this is another consequence of misalignment or the incorrect loadings of bearings. Rubbing between rotor and stator may also give rise to non-synchronous excitation and a wide range of phenomena.

Hence, the techniques presented in this chapter offer a 'first filter' to indicate changes in a machine's behaviour. To positively identify the nature of an event requires further insight into possible problems together with further analysis of the data.

Problems

2.1 Explain what is meant by 'order tracking' and why it is an important concept for the monitoring of rotating machines.

On shutdown, a large boiler feed pump runs down from 6000 rpm at an initial rate of 120 rpm/s. The response up to four times shaft speed is required.

   i. How would you sample to obtain maximum resolution, describing block size and sampling rate?

   ii. What is the resulting frequency resolution?

2.2 Assuming an exponential rundown rate in question one, how does the resolution vary during the rundown? If the rundown rate is linear, how is the resolution changed?

2.3 A signal is known to be $y = \sin(50\pi t) + 3\cos(100\pi t)$. Using 200 sampling points, obtain the FFTs using sample lengths of 0.5, 1, 2 and 4 s. Comment on the results.

2.4 A signal has the form $y = \sin(50\pi t) + 4\cos(55\pi t)$. Determine appropriate sampling rates and the number of samples to obtain adequate separation of the two components. Obtain the FFT and discuss your results.

2.5 A signal which is band limited in frequency can be expressed as

$$y(t) = \frac{1}{2\Delta\omega} \int_{\omega_0 - \Delta\omega}^{\omega_0 + \Delta\omega} \cos\omega t\, d\omega$$

where $2\Delta\omega$ is the total bandwidth. Calculate the time response and plot this function for values of $\Delta\omega = \pi$ and 50π, for a constant $\omega_0 = 100\pi$.

2.6 A scatter diagram contains 1000 data points which are described by the parameters x_i and y_i. It has been established that

$$\sum_{i=1}^{1000}(x_i - \bar{x})^2 = 73,570 \quad \sum_{i=1}^{1000}(x_i - \bar{x})(y_i - \bar{y}) = -23,600 \quad \sum_{i=1}^{1000}(y_i - \bar{y})^2 = 53,860$$

where $\bar{x} = -3$, $\bar{y} = 2$ are mean values. Derive the size and orientation of the ellipse which defines the standard deviation of this set of data. What is the probability of observing a point on the diagram within the interval $x = 7 \pm 1$, $y = 2 \pm \frac{1}{2}$?

References

Bossley, K.M., McKendrick, R.J., Harris, C.J. and Mercer, C., 1999, Hybrid computer order tracking, *Mechanical Systems and Signal Processing*, 13(4), 627–641.

Cooley, J.W. and Tukey, J.W., 1965, An algorithm for the calculation of complex Fourier series, *Mathematics of Computation*, 19, 297–301.

Friswell, M.I., Penny, J.E.T., Garvey, S.D. and Lees, A.W., 2010, *Dynamics of Rotating Machines*, Cambridge University Press, New York.

Fyfe, K.R. and Munck, E.D.S., 1997, Analysis of computed order tracking, *Mechanical Systems and Signal Processing*, 11(2), 187–205.

Irons, B.M. and Ahmad, S., 1980, *Finite Element Techniques*, Ellis-Horwood, Chichester, U.K.

Lees, A.W., Sinha, J.K. and Friswell, M.I., 2009, Model based identification of rotating machines, *Mechanical Systems and Signal Processing*, 23(6), 1884–1893.

Quiney, Z.A., 2011, Use of acoustic emission for bearing monitoring, PhD thesis, Swansea University, Swansea, U.K.

3

Modelling and Analysis

3.1 Introduction

In this chapter, consideration is given to some of the basic methods of analysis in the study of rotating machinery. In some respects, this is similar to the study of structural dynamics, but there is an important complicating factor: the variation of both the system and the forcing with shaft rotational speed. This is an obvious prerequisite to the material which follows in subsequent chapters.

We note that there are two fundamental distinctions between a structure and a rotating machine, as far as the dynamic behaviour is concerned. First, the dynamic properties of a rotating shaft depend on the rotational speed owing to gyroscopic terms and, in many cases, the property of the bearings. Second, the shaft rotation provides an energy source which implies that under certain conditions, vibrations can grow – a condition of instability. Both these issues will be addressed in this chapter.

3.2 Need for Models

The objective of any monitoring scheme on any kind of equipment is the rapid identification of any trend in performance. The overall problem can be approached at different levels, and one could devise a very simple system which equates increased vibration to a general deterioration and at some point removes the plant from service. However, with sophisticated plant, it is usually much more desirable to use the monitored data to gain an understanding of the machine loading and use this knowledge to guide maintenance and operations. This invariably requires some forms of mathematical model to aid understanding, and in most instances, for all except the simplest of machines, this will mean a finite element model (FEM).

Basically, understanding of the machine's behaviour is gained by comparing the measured vibration levels with those obtained from the model.

Of course this requires some care. A full discussion of the FEM is beyond the scope of the present text, but many books cover the basis of the approach (see, e.g. the classic text by Zienkievicz et al. 2005). The method as it applies to rotating machinery is discussed by Friswell et al. (2010), and a very brief outline is given in the following section.

3.3 Modelling Approaches

Why is this section included in a text on condition monitoring? Some engineers would regard modelling and plant measurements as two quite disparate disciplines, but if the effort is focussed on maximising the understanding of the internal processes of plant, then the two approaches are, in general, inextricably linked and are equally essential for a comprehensive status overview. A model without measurements is somewhat divorced from reality, whereas measurements in themselves do not give a view of the internal operation. In establishing the methodology of finite elements (FE), although consideration of a bending beam is among the simplest of examples, it is also one of the most important.

The rotor of many machines can be modelled as a rotating beam onto which one or more discs are attached. Some machines required more sophisticated models, particularly those in which significant distortion of the cross section takes place. Examples of this are in a minority and are discussed by Friswell et al. (2010).

3.3.1 Beam Models

FEM can be viewed as an extension to the Rayleigh–Ritz approach to the analysis of structures. In this approach, a trial function is taken for the body's displacement, and then the kinetic and potential energies can readily be calculated. The key point in seeking to represent motion in terms of trial function is the dramatic reduction in the number of degrees of freedom from a very large (effectively infinite) number to the parameters of the trial functions. Whereas in the classical Rayleigh–Ritz approach, only a few trial parameters were considered; in FEM considerably more are allowed. Fortunately, a considerable number of rotating machines can be adequately modelled using a simple linear element and so the principles of the FEM are illustrated for this case. Let us consider the steps in modelling a simple machine, the rotor of which can be represented as a beam mounted on bearings at each end. The first step is to divide the rotor into sections, called elements; in principle, we can choose the number of these elements, but this is a very important issue to which we will return. For the sake of illustration, we simply divide the rotor into two elements. At the end of each element,

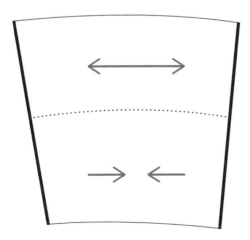

FIGURE 3.1
Strains during bending.

there is a node – which may be regarded as a point at which the displacement
is determined. The rotating assembly of many, but not all, rotating machines,
can be accurately represented by beams, and hence our attention is now
focussed on the modelling of beams. The simplest description of a bending
beam is that given by Euler. We consider a beam of rectangular cross section
(dimensions $a \times b$) and length L, which is considerably greater than either
a or b. It is clear, however, that there is a curved surface which can be drawn
through the beam on which there is no extension: this is called the neutral
axis, and in a uniform beam, it would lie along the centre of the beam. Points
within the beam above this surface are subject to tensile stresses, whilst
those below are subject to compression as shown in Figure 3.1. An important
assumption is that as the beam bends, cross sections remain plane and furthermore, they are at right angles to the neutral axis. In other words, there is
no shear deformation.

At some point, distance x from the neutral axis, the bending may be taken
to have a local radius of curvature R. The situation is shown in Figure 3.2,
and it is clear that all points above the neutral axis are in tension whilst
points below the neutral axis are in compression. Then, the strain is given by

$$\varepsilon = \frac{x}{R} \tag{3.1}$$

FIGURE 3.2
Simple rotor on two bearings.

And the corresponding stress is given as

$$\sigma = E\frac{x}{R} \tag{3.2}$$

Summing moments about the neutral axis gives

$$M_r = \int E\frac{x^2}{R}dx\,dy = \frac{EI}{R} \tag{3.3}$$

where
  I is the second moment of area
  E is Young's modulus

The radius of curvature R is the inverse of the second derivative of the bending deflection and hence

$$M_r = EI\frac{\partial^2 u}{\partial z^2} \tag{3.4}$$

Using this expression, the equation of motion of an Euler beam is readily derived. If we consider the motion of a small portion of the beam between z and $z + \delta z$, then by considering the difference in shear force

$$P(z+\delta z) - P(z) = \rho A\frac{d^2 u}{dt^2} \tag{3.5}$$

Since the shear force is the derivative of the bending moment, M_r, the Euler–Bernoulli beam equation becomes

$$\frac{\partial^2}{\partial z^2}\left(EI\frac{\partial^2 u}{\partial z^2}\right) - \rho A\frac{d^2 u}{dt^2} = 0 \tag{3.6}$$

In fact from the Euler–Bernoulli beam equation, the nth natural frequency for this case is given by (using the usual notation)

$$\omega_n = \frac{n^2\pi^2}{L^2}\sqrt{\frac{EI}{\rho A}} \tag{3.7}$$

where
  L is length
  A is cross sectional area
  ρ is the density

3.3.2 Finite Element Method

It is important to recall that the FE approach is an application of the Rayleigh–Ritz variational method, albeit a highly sophisticated extension. Just as with all such approaches, in calculating natural frequencies, accuracy improves when the trial function is given sufficient freedom to faithfully represent the shape of the modes of the machine. This implies, in general, that the more the elements in the model, the better the result, but of course, a significant penalty is paid for this in terms of computing effort. In practice, one must also ask what accuracy is required. In dealing with the resolution of problems on real machines, the analyst is usually presented with the problem of reconciling measured data with a theoretical model in an attempt to gain understanding and insight. Clearly, the accuracy must be sufficient to aid this process, but there is a point at which improved precision is largely academic.

In fact, the idealisation of a given rotor is the area where the analyst's skill is most important. There are two major and sometimes conflicting assessments to be made. On the one hand, the model must reflect the geometry of the machine, with appropriate changes of section, but consideration is also needed as to the number of elements required to give the appropriate measure of confidence. It is shown that these two factors are sometimes reconciled by using an approach known as 'node condensation' whereby the geometry is modelled in considerable detail, but then the resulting very large model is reduced in size prior to the vibration calculations. We first consider the problem of modelling a uniform beam which is pinned at each end. Initially, a Euler beam is considered since an analytic solution is readily available. The corrections due to shear and rotary inertia are considered subsequently.

3.3.2.1 Element Formulation

Consider a beam element with nodes at each end lying along the z-axis. At each node, it is assumed that the displacement u transverse to its axis and the slope du/dz are nodal parameters. Assume that the transverse displacement within the element is given by

$$u = a_0 + a_1 z + a_2 z^2 + a_3 z^3 \tag{3.8}$$

Because the parameters a_n are not readily related to the global system, they should be eliminated and this can be done using nodal variables. Note here, that for this element, there are now only four degrees of freedom, corresponding to the parameters a_0, a_1, a_2 and a_3. We may form expressions for the four nodal variables and write in vector and matrix notation for an element of length l_e:

$$
\begin{Bmatrix} u_1 \\ \left(\dfrac{du}{dz}\right)_1 \\ u_2 \\ \left(\dfrac{du}{dz}\right)_2 \end{Bmatrix}
=
\begin{bmatrix} 1 & 0 & 0 & 0 \\ 0 & 1 & 0 & 0 \\ 1 & l_e & l_e^{\,2} & l_e^{\,3} \\ 0 & 1 & 2l_e & 3l_e^{\,2} \end{bmatrix}
\begin{Bmatrix} a_0 \\ a_1 \\ a_2 \\ a_3 \end{Bmatrix}
\tag{3.9}
$$

The subscripted values on the left-hand side of the equation refer to values at the respective nodes. From this equation, we may readily write the vector of the $\{a\}$ parameters in the form

$$
\{a\} = [C]^{-1}\{v\}
\tag{3.10}
$$

where $\{v\}^{T} = \left\{ u_1 \left(\dfrac{du}{dz}\right)_1 \; u_2 \left(\dfrac{du}{dz}\right)_2 \right\}$ and it is readily shown that

$$
[C]^{-1} =
\begin{bmatrix}
1 & 0 & 0 & 0 \\
0 & 1 & 0 & 0 \\
-3/l_e^{\,2} & -2/l_e & 3/l_e^{\,2} & -1/l_e \\
2/l_e^{\,3} & 1/l_e^{\,2} & -2/l_e^{\,3} & 1/l_e^{\,2}
\end{bmatrix}
\tag{3.11}
$$

To calculate the potential energy in the element during deformation in the prescribed form, we may write

$$
PE = \int_0^{l_e} EI \left(\frac{\partial^2 u}{\partial z^2}\right)^2 dz
\tag{3.12}
$$

which may be expressed as

$$
PE = EI \int_0^{l_e} (2a_2 + 6a_3 z)^2 \, dz
\tag{3.13}
$$

However, because we wish to eliminate the $\{a\}$ parameters, it is far more convenient to express the energy in matrix terms, hence

$$
PE = EI \int_0^{l_e} \{a\}^{T}
\begin{Bmatrix} 0 \\ 0 \\ 2 \\ 6z \end{Bmatrix}
\begin{Bmatrix} 0 & 0 & 2 & 6z \end{Bmatrix} \{a\} \, dz
\tag{3.14}
$$

but since we need to express this energy in terms of the nodal variables, the $\{a\}$ parameters are eliminated and we obtain from Equation 3.10

$$PE = EI \int_0^{l_e} \{v\}^T [\mathbf{C}]^{-T} \begin{Bmatrix} 0 \\ 0 \\ 2 \\ 6z \end{Bmatrix} \{0 \quad 0 \quad 2 \quad 6z\} [\mathbf{C}]^{-1} \{v\} dz \tag{3.15}$$

After completing the integrals and multiplications, an expression for the potential energy of the element is obtained. This becomes

$$PE = \{v\}^T [\mathbf{K}]\{v\}$$

where

$$\mathbf{K} = \frac{2EI}{l_e^3} \begin{bmatrix} 6 & 3l_e & -6 & 3l_e \\ 3l_e & 2l_e^2 & -3l_e & l_e^2 \\ -6 & -3l_e & 6 & -3l_e \\ 3l_e & l_e^2 & -3l_e & 2l_e^2 \end{bmatrix} \tag{3.16}$$

The mass matrix is calculated by considering the kinetic energy. In this case, we obtain

$$KE = \int_0^{l_e} \{v\}^T [\mathbf{C}]^{-T} \{1 \quad z \quad z^2 \quad z^3\} \begin{Bmatrix} 1 \\ z \\ z^2 \\ z^3 \end{Bmatrix} [\mathbf{C}]^{-1} \{v\} dz \tag{3.17}$$

On completing the calculations, the expression for kinetic energy is given by

$$KE = \omega^2 \{v\}^T [\mathbf{M}]\{v\}$$

where

$$\mathbf{M} = \frac{ml_e}{420} \begin{bmatrix} 156 & 22l_e & 54 & -13l_e \\ 22l_e & 4l_e^2 & 13l_e & -3l_e^2 \\ 54 & 13l_e & 156 & -22l_e \\ -13l_e & -3l_e^2 & -22l_e & 4l_e^2 \end{bmatrix} \tag{3.18}$$

where m is the mass per unit length of the beam.

However, there is one further step in deriving the element matrices. So far, we have analysed the motion in a single direction, whereas in most problems on machinery, we must consider at least two orthogonal motions: this is easily accomplished and the element matrices become 8×8 rather than 4×4. Using the Euler–Bernoulli formulation for an axisymmetric rotor, the element mass matrix becomes

$$
\mathbf{M_e} = \frac{\rho_e A_e l_e}{420}
\begin{bmatrix}
156 & 0 & 0 & 22l_e & 54 & 0 & 0 & -13l_e \\
0 & 156 & -22l_e & 0 & 0 & 54 & -13l_e & 0 \\
0 & -22l_e & 4l_e^2 & 0 & 0 & -13l_e & -3l_e^2 & 0 \\
22l_e & 0 & 0 & 4l_e^2 & 13l_e & 0 & 0 & -3l_e^2 \\
54 & 0 & 0 & 13l_e & 156 & 0 & 0 & -22l \\
0 & 54 & -13l_e & 0 & 0 & 156 & 22l_e & 0 \\
0 & -13l_e & -3l_e^2 & 0 & 0 & 22l_e & 4l_e^2 & 0 \\
-13l_e & 0 & 0 & -3l_e^2 & -22l_e & 0 & 0 & 4l_e^2
\end{bmatrix}
\tag{3.19}
$$

The element stiffness matrix becomes

$$
\mathbf{K_e} = \frac{E_e I_e}{l_e^3}
\begin{bmatrix}
12 & 0 & 0 & 6l_e & -12 & 0 & 0 & 6l_e \\
0 & 12 & -6l_e & 0 & 0 & -12 & -6l_e & 0 \\
0 & -6l_e & 4l_e^2 & 0 & 0 & 6l_e & 2l_e^2 & 0 \\
6l_e & 0 & 0 & 4l_e^2 & -6l_e & 0 & 0 & 2l_e^2 \\
-12 & 0 & 0 & -6l_e & 12 & 0 & 0 & -6l_e \\
0 & -12 & 6l_e & 0 & 0 & 12 & 6l_e & 0 \\
0 & -6l_e & 2l_e^2 & 0 & 0 & 6l_e & 4l_e^2 & 0 \\
6l_e & 0 & 0 & 2l_e^2 & -6l_e & 0 & 0 & 4l_e^2
\end{bmatrix}
\tag{3.20}
$$

The simple element derived here is based on Euler–Bernoulli beam theory. Whilst this is acceptable for the modelling of slender rotors (with a high length/diameter ratio), it is often necessary to base the model on the more sophisticated Timoshenko beam theory which includes shear and rotary inertia effects. After a slightly more complicated derivation, the stiffness becomes

$$
\mathbf{K} = \frac{EI}{(1 + \Phi_e)l_e^2}
\begin{bmatrix}
12 & 6l_e & -12 & 6l_e \\
6l_e & l_e^2(4 + \Phi_e) & -6l_e & l_e^2(2 - \Phi_e) \\
-12 & -6l_e & 12 & -6l_e \\
6l_e & l_e^2(2 - \Phi_e) & -6l_e & l_e^2(4 + \Phi_e)
\end{bmatrix}
\tag{3.21}
$$

where

$$\Phi_e = \frac{12EI}{\kappa GA_e l_e^2}$$

κ is the shear constant
G is the shear modulus
A_e is the cross section area

The derivation of the mass matrix is rather tedious, but it produces two terms which are each 4×4 matrices representing the mass and rotary inertia, respectively. Full details can be found in Chapter 4 of Friswell et al. (2010). The mass matrix for the element is now given as

$$\mathbf{M_e} = \frac{\rho_e A_e l_e}{840(1+\Phi_e)^2} \begin{bmatrix} m_1 & m_2 & m_3 & m_4 \\ m_2 & m_5 & -m_4 & m_6 \\ m_3 & -m_4 & m_1 & -m_2 \\ m_4 & m_6 & -m_2 & m_5 \end{bmatrix}$$

$$+ \frac{\rho_e I_e}{30(1+\Phi_e)^2 l_e} \begin{bmatrix} m_7 & m_8 & -m_7 & m_8 \\ m_8 & m_9 & -m_8 & m_{10} \\ -m_7 & -m_8 & m_7 & -m_8 \\ m_8 & m_{10} & -m_8 & m_9 \end{bmatrix}$$ (3.22)

where

$$m_1 = 312 + 588\Phi_e + 288\Phi_e^2 \qquad m_6 = -\left(6 + 14\Phi_e + 7\Phi_e^2\right)l_e^2$$

$$m_2 = \left(44 + 77\Phi_e + 35\Phi_e^2\right)l_e \qquad m_7 = 36$$

$$m_3 = 108 + 252\Phi_e + 140\Phi_e^2 \qquad m_8 = \left(3 - 15\Phi_e\right)l_e$$

$$m_4 = -\left(26 + 63\Phi_e + 35\Phi_e^2\right)l_e \quad m_9 = \left(4 + 15\Phi_e + 10\Phi_e^2\right)l_e^2$$

$$m_5 = \left(8 + 14\Phi_e + 7\Phi_e^2\right)l_e^2 \qquad m_{10} = -\left(1 + 5\Phi_e - 5\Phi_e^2\right)l_e^2$$

The second part of the mass matrix represents the effect of rotary inertia. It should be emphasized that although the derivation of the Timoshenko element is somewhat more complicated than that of the Euler element, there is no added difficulty in use. This is discussed in Section 3.3.3.

3.3.2.2 Matrix Assembly

Both components of the energy of the element have now been written in terms of the nodal coordinates, and then the elements may then be assembled into structures. To illustrate this, consider the beam comprising two elements in Figure 3.2. Consider the vibration in one plane only and there are now three nodes with six degrees of freedom. Denoting elements from the nth element as $(k_{ij})_n$, then the assembled stiffness matrix for the complete beam will be

$$\mathbf{K} = \begin{bmatrix} (k_{11})_1 & (k_{12})_1 & (k_{13})_1 & (k_{14})_1 & 0 & 0 \\ (k_{21})_1 & (k_{22})_1 & (k_{23})_1 & (k_{24})_1 & 0 & 0 \\ (k_{31})_1 & (k_{32})_1 & (k_{33})_1 + (k_{11})_2 & (k_{34})_1 + (k_{12})_2 & (k_{13})_2 & (k_{14})_2 \\ (k_{41})_1 & (k_{42})_1 & (k_{43})_1 + (k_{21})_2 & (k_{44})_1 + (k_{22})_2 & (k_{23})_2 & (k_{24})_2 \\ 0 & 0 & (k_{31})_2 & (k_{32})_2 & (k_{33})_2 & (k_{34})_2 \\ 0 & 0 & (k_{41})_2 & (k_{42})_2 & (k_{43})_2 & (k_{44})_2 \end{bmatrix} \quad (3.23)$$

and the overall mass matrix is made up in the same manner. Thus, we can express the energy, and therefore, the force distribution in terms of the nodal parameters of the full structure.

For more complicated element types, the mathematical approach is a little different, but the physical basis of the argument remains the same.

The system matrices are assembled following the pattern shown in Equation 3.23, adding in terms for masses on the shaft and bearing parameters. The an equation of motion for the system can be given as

$$\mathbf{Kx} + \mathbf{C\dot{x}} + \mathbf{G\Omega\dot{x}} + \mathbf{M\ddot{x}} = \mathbf{F}(t) \quad (3.24)$$

Two extra terms have been added to this equation: $\mathbf{C}$ represents the damping, whilst $\mathbf{G}$ is the gyroscopic term. The damping is sometimes difficult to quantify accurately but arises, in the most part, from the bearings. Damping arising in the rotor itself demands a rather different treatment, and this can cause instabilities (see Friswell et al. 2010). The gyroscopic term arises from consideration of angular momentum conservation and the matrix $\mathbf{G}$ is skew-symmetric: the important point to note is that the presence of this term makes the dynamic properties of the system dependent on the rotational speed of the shaft. The importance of this term clearly depends on the geometry of the shaft under consideration; in practical machines, gyroscopic terms are rarely important for cases in which the dominant inertias are between the bearings, but for overhung rotors they become a crucial consideration.

It is worthwhile reflecting on what has been achieved by representing the machine in this manner. Despite considerable underlying algebra, the physical importance of the process is quite straightforward. By imposing a fixed functional relationship within each element (Equation 3.8), the original system with an infinite number of degrees of freedom has been reduced to one with a finite number of degrees of freedom (equal in total to four times the number of nodes). The degree to which the motion of the model accurately represents the behaviour of the real system is determined by the ability of the model to represent the true deflection, and this will be determined by the number of elements in the model. In principle, clearly, the more elements the better, but this approach leads to heavy computational overheads so it is important to gain some insight into the way in which model refinement is reflected in accuracy.

To illustrate the point, we consider the following example:

Example 3.1

A circular shaft with a diameter of 10 mm is supported on bearings at either end, 0.5 m apart. Determine the first four natural frequencies using 1, 2, 4 and 8 Euler–Bernoulli FEs.

Solution

Since the model is using Euler–Bernoulli beam theory, an exact solution is available. This is given by

$$\omega_n = n^2 \pi^2 \sqrt{\frac{EI}{ml^4}}$$

The results of the FE calculations are shown in Table 3.1. As shown in the table, the lower order modes are predicted well with few elements, but the higher modes require a much finer FE mesh. This is simply a manifestation of the need to represent the mode's deflection shape.

Figure 3.3 shows the model of this simple rotor with 1, 2, 4 and 8 elements, respectively.

There are several points which are clear from this fairly simple set of calculations:

a. The FE-derived natural frequency is, in all cases, higher than the exact value. This is always the case with this type of element (which is said to be conforming meaning that there is continuity of the displacement and slope between adjoining

TABLE 3.1

Values of Natural Frequencies (Hz) for Simply Supported Beam

No. of Elements	1	2	4	8	Exact
Mode 1	90.5	81.8	81.5	81.5	81.5
Mode 2	414.6	361.9	327.3	326.1	326.0
Mode 3	–	909.5	746.9	734.5	733.5
Mode 4	–	1658	1447	1309	1304

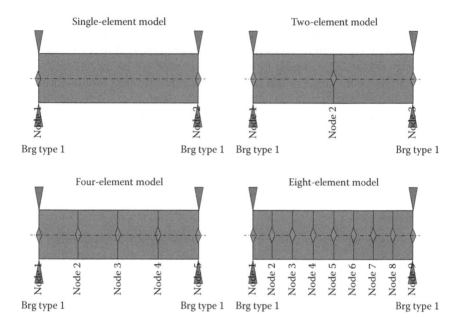

FIGURE 3.3
Rotor modelling with FE.

          elements). The reason for this is clear physically: the model
          is more restricted than the true plant and consequently is
          stiffer.
     b. The accuracy increases with the number of elements but
          decreases with mode number.
     c. A model with only two elements gives the first mode accurately
          but shows considerable error for the second mode.

Having set out the basic framework for the modelling of rotating machines,
we now consider the evaluation of measured data. Often modelling and the
plant data are considered as quite separate entities, but throughout the present text, it is proposed that they should be intimately related: experience
on the plant should inform the modelling whilst predictions of the model
aid understanding of the plant. The two are inseparable. Whilst an analytic approach is suitable for the study of an ideal uniform beam, it becomes
more clumsy to apply Equation 3.6 to a beam of non-uniform cross section.
For the study of most realistic analyses, the most convenient approach is
with the FEM.

3.3.3 Modelling Choices

Of course, the examples so far all use the Euler beam theory which neglects
shear and rotary inertia. The use of this type of element is just one of the

TABLE 3.2

Influence of Aspect Ratio on Normalised Frequency Calculations

	Ratios (D/L)				
Mode	0.025	0.05	0.075	0.1	Euler Solution
1	0.9992	0.9970	0.9933	0.9882	1
2	3.9881	3.9531	3.8971	3.8235	4
3	8.9414	8.7717	8.5136	8.1947	9
4	15.8213	15.3161	14.5915	13.7567	16

choices that the engineer must make on how to represent the system. The main decisions to be made are the following:

a. Is a beam model appropriate?
b. Is shear deformation and/or rotary inertia important (depending on the rotor's aspect ratio and the modes considered)?
c. The number of elements used.
d. Other factors such as bearings, seals and the distribution of mass.

Timoshenko's beam theory includes both of these effects, and Table 3.2 shows the difference these terms make for a circular cross section using 16 elements. This table is expressed in terms of the normalised frequency, f/Ξ_0 where

$$\Xi_0 = \frac{L^2}{\pi^2}\sqrt{\frac{\rho A}{EI}} \tag{3.25}$$

The table shows the normalised frequency or a number of aspect ratios. Note how the significance of the extra terms increases for higher mode numbers.

Whilst Timoshenko's model is superior to that of Euler for thick beams, it is not exact, and as the ratio of diameter to length increases, the whole idealisation of beams breaks down and the only adequate model would be a full 3D solid model. Fortunately, this is rarely necessary for the analysis of rotating machines. Although the derivation is a little involved, this element is available in all commercial software, and it is no more difficult to use than the Euler version. It is, therefore, advisable to use this as standard in rotor calculations except, of course, where more sophisticated models are required.

3.4 Analysis Methods

In the previous chapter, a description is given of the main ways of presenting measured data from rotating plant. There is a considerable quantity of data available in many instances, and sometimes, a rather confusing scenario is portrayed: it is often necessary to view data in a number of different ways in order to digest the significance. In any case of fault diagnosis, the essential feature is to correlate the observation with an understanding of the plant operation, and this means some theoretical model. On occasions, the required model may be a rudimentary idealisation of the real machine, but with increasing plant sophistication, a detailed model will often be needed, and the FEM provides the most important route to establish these models. In many instances, there is no sufficient information to provide an accurate model, but this does not negate the usefulness of the modelling approach in gaining understanding. Chapter 7 discusses recent techniques for the correlation of models with plant measurements, and in particular, we discuss the derivation of the properties of supporting structures which is a common difficulty, particularly in the case of very large machines. Once established, the model can produce a range of presentations of results. In some cases, these replicate the measured plots discussed in the previous section, but since there is greater access to data, some other equally informative plots are available.

The development of FEM has transformed the analysis, design and understanding of rotating machinery. It provides a general formulation to gain insight into the dynamics of machinery.

3.4.1 Imbalance Response

Mass unbalance is the most prevalent fault in turbo-machinery so it is natural to begin our discussion with the assessment of an unbalanced shaft. This topic, arguably the most important in the study of rotating machinery, is discussed in detail in Chapter 4; it is, however, convenient to give a short summary at this point in order to utilise some of the basic ideas in subsequent discussions. The general equation of motion Equation 3.24 may be rewritten to give the response at a shaft speed corresponding to Ω radians per second. Then, the response x is given by

$$\mathbf{K}x + j\mathbf{C}\Omega x + j\mathbf{G}\Omega^2 x - \mathbf{M}\Omega^2 x = \mathbf{M}\Omega^2 e \qquad (3.26)$$

The important point to note here is that the shaft itself is providing the forcing term which gives rise to vibration. The solution to this may be conveniently computed as

$$x = [\mathbf{K} + j\mathbf{C}\Omega + j\mathbf{G}\Omega^2 - \mathbf{M}\Omega^2]^{-1}\mathbf{M}\Omega^2 e \qquad (3.27)$$

This expression can be used to calculate and plot a rundown curve given a vector of mass eccentricities. It should be emphasised that this calculation is quasi-static in the sense that it strictly simulates a rundown taken over an infinite length of time. However, the difference caused by this should be small.

Before leaving Equation 3.27, we note that for even moderate size models over a frequency range, this represents a substantial computation. For reasons of computational efficiency, the calculation is normally carried out in terms of modal parameters. This approach is explained in Section 3.5.1.

3.4.2 Campbell Diagram

Following the philosophy for non-rotating structures, the first step here is to consider natural frequencies, but with the added complication that these natural frequencies will be a function of rotational speed. A plot of the natural frequencies as a function of shaft speed is the Campbell diagram, an example of which is shown in Figure 3.4.

For a given rotation speed Ω, the natural frequencies are given by

$$\mathbf{Kx} + j\mathbf{C}\omega\mathbf{x} + j\mathbf{G}\Omega\omega\mathbf{x} - \mathbf{M}\omega^2\mathbf{x} = 0 \tag{3.28}$$

In the case Ω and ω are distinct quantities, the former being the rotation speed whilst the latter is the frequency of vibration. Because the exciting forces are

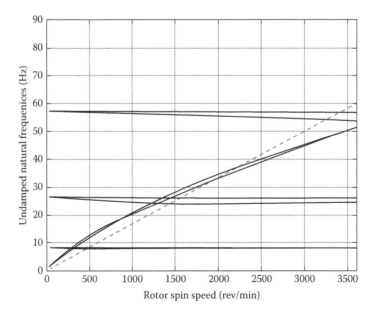

FIGURE 3.4
Campbell diagram.

usually locked to the shaft rotation, lines along which the frequency is in a fixed relationship to the rotation speed are of particular importance, and the intersections of these lines and the natural frequency lines are called critical speeds. These speeds are of crucial importance in both the design and operation of rotating machinery. But even this does not give the whole picture of what is a very complicated story: in particular, the presence of a critical speed on a Campbell diagram does not necessarily imply a problem, because the particular resonance may not be excited to any significant degree. A resonance may not be excited owing to heavy damping, inappropriate distribution of forces or geometric effects; a very important example of this occurs in the case of reverse modes which cannot be excited by unbalance in the case of a machine mounted on symmetrical supports. This is discussed further in Chapter 4. Whilst the Campbell diagram is invaluable as a diagnostic aid, owing to the lack of information on damping, it is incomplete and we seek to supplement our system knowledge.

The four plots shown in Figure 3.5a through 3.5d give an overview of some of the varieties seen in Campbell diagrams. The four cases shown all relate to a simple fan with a large impeller mounted on the shaft between rolling element bearings. Because of the large impeller, gyroscopic moments play a part in some of the modes depending on the angular deflection of the mode in question. In Figure 3.5a since the impeller is centrally mounted, the first mode is just a 'bounce' motion; hence, there is no gyroscopic splitting. There is splitting in the second mode. Note that in this case, there are only two critical speeds (one forward and one reverse). In Figure 3.5b, the supporting structure has been given different stiffness in the x and y direction, and this can be seen by the frequency splits at zero shaft speed. In Figure 3.5c and d, the axial location of the impeller has been changed and is off-centre. This means that even the lowest mode of vibration will involve some rotation, and hence there is gyroscopic splitting of the first mode. Note also that with simple bearings (i.e. not oil film type), it is easily shown that forward modes increase in frequency with rotation speed whereas reverse modes decrease in frequency.

3.4.3 Analysis of Damped Systems

In the discussion so far, the evaluation of rotor response as a function of rotor speed has been demonstrated for simple cases, and a more extensive survey of the topic is given in Chapter 4. Although the forced response is often the most valuable aspect of analysis, understanding the natural frequencies and mode shapes will often add significantly to the engineer's insight into machine operation. In analysing the free vibration, the analyst seeks the damping coefficient of each mode in addition to the natural frequencies and mode shapes.

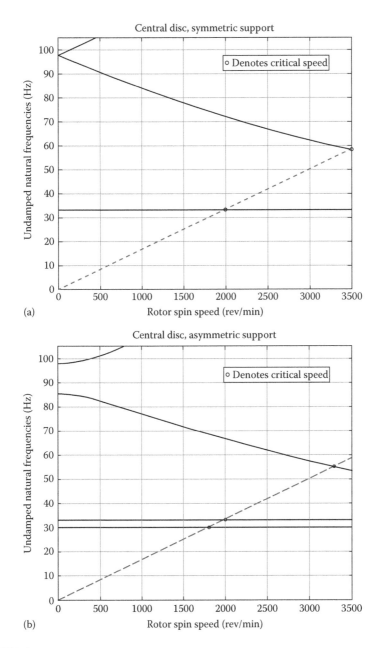

(a)

(b)

FIGURE 3.5
(a–d) Campbell diagrams for various geometries. (*Continued*)

FIGURE 3.5 (Continued)
(a–d) Campbell diagrams for various geometries.

At first sight, this presents a problem, as there are now three matrices to deal with in the analysis of the form (neglecting the Gyroscopic term)

$$\mathbf{Kx} + \mathbf{C\dot{x}} + \mathbf{M\ddot{x}} = 0 \tag{3.29}$$

A method is required to obtain natural frequencies, mode shapes and damping coefficients with the three matrices K, M and C involved in the calculation.

This problem is most conveniently solved by expressing Equation 3.29 in term of so-called 'state space'. To do this, we regard x and $\dot{x}$ as two separate variables describing position and velocity, respectively. This means that there must be an extra equation expressing the fact that velocity is the time derivative of position, and so the second order Equation 3.29 becomes two first-order equations

$$\mathbf{K}x + \mathbf{C}\dot{x} + \mathbf{M}\frac{d\dot{x}}{dt} = 0 \tag{3.30}$$

$$\dot{x} = \frac{dx}{dt} \tag{3.31}$$

These two equations may now be combined into a single one of the form

$$\begin{bmatrix} \mathbf{K} & 0 \\ 0 & -\mathbf{M} \end{bmatrix} \begin{Bmatrix} x \\ \dot{x} \end{Bmatrix} + \begin{bmatrix} \mathbf{C} & \mathbf{M} \\ \mathbf{M} & 0 \end{bmatrix} \frac{d}{dt} \begin{Bmatrix} x \\ \dot{x} \end{Bmatrix} = 0 \tag{3.32}$$

This is of the form

$$\mathbf{A}q + \mathbf{B}\dot{q} = 0 \tag{3.33}$$

The unknown vector is now $\{x \quad \dot{x}\}^T$ which clearly has twice the dimension of the original problem. Note also that this analysis gives a direct calculation of natural frequency, rather than the natural frequency squared obtained in the un-damped second-order calculation. In general, the natural frequencies will be complex, of the form $\omega_k = \alpha_k + j\beta_k$, and it is worth reflecting on the physical meaning of this. Recall that an equation such as Equation 3.33 is solved by assuming a form $\mathbf{q} = \mathbf{q}_0 e^{\lambda t}$ and after substitution, this leads to the eigenvalue equation

$$\mathbf{A}q_0 + \lambda \mathbf{B}q_0 = 0 \tag{3.34}$$

If there are n degrees of freedom in the original (second order) problem, there will be $2n$ solutions to the eigenvalue problem of Equation 3.32, each comprising an eigenvalue, λ, and corresponding eigenvector, $\mathbf{q}_0$. The kth solution to the free vibration Equation 3.33 becomes

$$q_k = q_{0k} e^{\alpha_k t}(A \sin \beta_k t + B \cos \beta_k t) \tag{3.35}$$

It is clear from this that, provided α is negative, any disturbance will decay in time, but if α is greater than zero, disturbance will grow and the system will be unstable. *A* and *B* are just constants chosen to fit the initial conditions. By comparison with a single degree-of-freedom system, it is easily shown that the *k*th natural frequency ω_{nk} and the damping coefficient ζ_k are given by

$$\omega_{nk} = \sqrt{\alpha_k^2 + \beta_k^2} \tag{3.36}$$

$$\zeta_k = \frac{\alpha_k}{\sqrt{\alpha_k^2 + \beta_k^2}} \tag{3.37}$$

It is clear from these two relationships that β_k gives the damped natural frequency and that

$$\omega_{dk} = \beta_k = \omega_{nk}\sqrt{1 - \zeta_k^2} \tag{3.38}$$

To clarify these issues, consider the equation of motion for the free vibration of a single degree-of-freedom system. This is described by

$$kx + c\dot{x} + m\ddot{x} = 0 \tag{3.39}$$

Assuming the trial form $x = x_0 e^{st}$ leads to the equation

$$(k + cs + ms^2)x_0 e^{st} = 0 \tag{3.40}$$

From this, the two roots are given as

$$s = \frac{-c \pm \sqrt{c^2 - 4km}}{2m} = \frac{-c}{2m} \pm \sqrt{\left(\frac{c}{2m}\right)^2 - \left(\frac{k}{m}\right)^2} \tag{3.41}$$

Setting $\zeta = (c/2m)\sqrt{m/k}$ and taking a factor of (-1) out of the square root gives

$$s = -\zeta\omega_n \pm j\omega_n\sqrt{1 - \zeta^2} \tag{3.42}$$

Note that when $\zeta = 1$, the imaginary part of s, corresponding to the oscillatory component, vanishes and the solution becomes real. Under this condition, the system is critically damped and all oscillation ceases. As an example, Figure 3.6 uses a model of an alternator and turbine on four bearings and shows a plot of the damping coefficient, ζ_k against the natural frequency, ω_{nk}

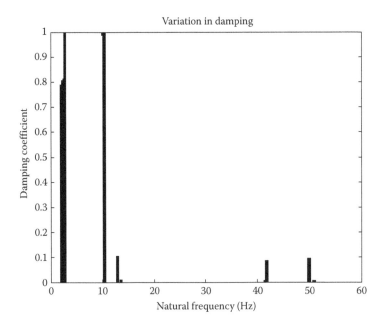

FIGURE 3.6
Damping coefficients for different modes.

for the first few modes. For this particular model, variation with rotor speed is not significant but could be readily included in a 3D plot.

In a multi-degree-of-freedom system, this applies to individual modes; although in many rotor systems, overall damping may be modest, but specific modes may be critically damped or overdamped. This type of behaviour is particularly prevalent in systems with oil-journal bearings (which usually account for an overwhelming proportion of the system damping).

There is one further issue related to system damping and the resulting dynamic behaviour. In the analysis of structures, it is frequently assumed that the damping matrix is a linear combination of the stiffness and mass matrices, that is $\mathbf{C} = a\mathbf{K} + b\mathbf{M}$.

This is the so-called 'classical' or 'proportional' model of damping. Various authors have sought some justification for such a model but in truth there is none. The reason this model is used is that it leads to substantial mathematical simplification and for modest damping at least, the answers are reasonable. This is the case because, in many cases, the damping is only important in parts of the frequency range close to a resonance. In many rotating machines, however, the damping is dominated by the bearings and this has some interesting consequences.

An important feature of 'proportional damping' is that the mode shapes are real and are those of the un-damped system. More general damping leads to complex eigen-functions (mode shapes), and we now consider what this means physically. Suppose that the kth mode shape takes the form

$$\psi_k = A_k(x) + jB_k(x) \tag{3.43}$$

Multiplying this by the time dependence, taking the real part yields

$$y_k = \text{Re}((A_k(x) + jB_k(x))e^{\alpha_k \omega_{nk} t}(\cos \omega_{nd} t + j \sin \omega_{nd} t)) \tag{3.44}$$

Hence for this modal contribution, we get

$$y_k = e^{\alpha_k \omega_{nk} t}(A_k(x) \cos \omega_{dk} t - B_k(x) \sin \omega_{dk} t) \tag{3.45}$$

The exponential term will, of course, settle owing to repeated forcing, but the interesting point to note is the phase of this component given by

$$\theta_k = \tan^{-1}\left(\frac{-B_k(x)}{A_k(x)}\right) \tag{3.46}$$

This varies with x and so we see that the complex mode represents a travelling wave in the machine, rather than a stationary wave represented by a real mode shape.

In any case, on an actual machine, the response will comprise contributions from a number of modes, and although these modes may be either real or complex, their summation will produce phases that vary with location.

3.4.4 Root Locus and Stability

The previous section discusses the treatment of the damping term in the equation of motion. The role of damping is crucial in rotating machinery, and this is another area where the situation is much more complicated than the dynamics of non-rotating structures. The fundamental problem is that machines have an energy input from the rotation of the shaft, and under certain circumstances, some of this energy can be transferred into the modes of vibration, and the result is that minor disturbances grow in time rather than decaying. There are three potential causes of rotor instability, which are as follows:

- Inappropriate bearing design or loadings
- High damping on the rotor
- Excessive rotor asymmetry

Within the context of fault diagnosis, the first of the alternatives will be the one most commonly met as alignment changes alter the static load on multi-bearing machines. This may manifest itself as a sub-synchronous whirl, but this will be discussed more fully in Chapter 6. The essential point to note here is that instability is present when the real part of one of the eigenvalues becomes positive. (The implications of this are explained shortly.) In any case, the damping is clearly an import parameter as it plays a crucial role in determining the practical importance of a mode.

Whilst a Campbell diagram is an extremely useful tool in the study of machinery, it gives incomplete information. Whilst it is important to be aware of the existence of natural frequencies, this is only a part of the story and a mode may either be heavily damped or may not be excited owing to the nature of the forcing function. In any event, some knowledge of the damping is clearly necessary.

In the root-locus diagram, the complex frequency of each mode is plotted on the Argand plane. The term 'root locus' for this plot is now widely accepted, but this is a little unfortunate as it implies a (non-existent) relationship to the analysis method of the same name in control theory. Figure 3.7 gives an example of a root-locus diagram. Each line shows the trajectory of a modal frequency as shaft rotational speed varies over the range. The *y*-coordinate represents the frequency of oscillation whilst the *x*-coordinate gives the decay rate (and from this the damping coefficient is readily determined). Any root showing in the right-hand side of the diagram grows with time and leads to an instability.

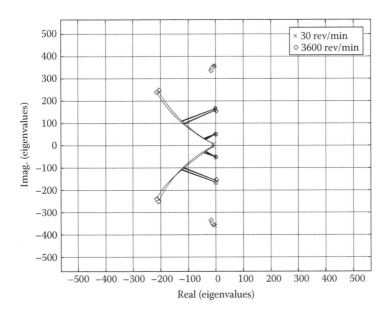

FIGURE 3.7
Root locus plot.

3.4.5 Overall Response

The root-locus plot is a useful tool for understanding stability issues, but it is only at best a very rough guide in assessing the importance of a particular mode in the dynamics of an overall system. To arrive at something potentially more helpful, we consider the forced response to a harmonic excitation. From Equation 3.24, observe that

$$\mathbf{x} = [\mathbf{K} + j\mathbf{C}\omega + j\mathbf{G}\Omega\omega - \mathbf{M}\omega^2]^{-1} f(\omega) \tag{3.47}$$

From which it can be inferred that an important parameter is

$$D = \det(\mathbf{K} + j\mathbf{C}\omega + j\mathbf{G}\Omega\omega - \mathbf{M}\omega^2) \tag{3.48}$$

Figure 3.8 shows a plot of the reciprocal of the determinant of the dynamic stiffness, as in Equation 3.46. In evaluating this, care is needed in scaling to avoid numerical difficulties.

Whilst this cannot be regarded as a precise measure of response, it does give an indication of potential problems without detailed consideration of forcing positions.

These 3D plots can help give insight into the relative importance of different resonance conditions. However, it is important to appreciate that even this does not present the whole story: in particular, it says nothing about stability. To understand this issue, we need to reflect on the nature of second-order

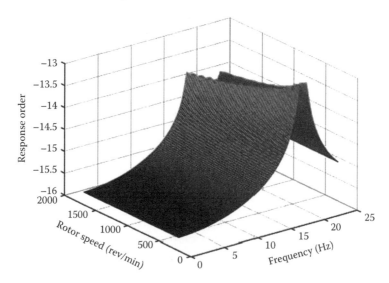

FIGURE 3.8
Plot of the reciprocal of the determinant (Equation 3.46).

differential equations of the type which describe the behaviour of rotating machines. In non-dimensional form, we may write for a single degree-of-freedom system

$$\ddot{x} + 2\zeta\omega_0\dot{x} + \omega_0^2 x = \frac{F}{m} \qquad \qquad (3.49)$$

Let the force, F, arise from imbalance and take the form $F = m\varepsilon\Omega^2 \cos \Omega t$, where ε is the mass eccentricity, then the full solution of the equation becomes

$$x = \frac{\varepsilon\Omega^2 \cos \Omega t}{\left(-\omega^2 + 2j\zeta\omega\omega_0 + \omega_0^2\right)} + A \exp\left(-\zeta\omega_0 t\right)\cos\left(\omega_0\sqrt{1-\zeta^2}t + \phi\right) \qquad (3.50)$$

In this expression, the constants A, ϕ are determined by the initial conditions. The first term gives the steady-state response: this gives the response to the force but it tells us nothing about the stability of the system. On the other hand, the second term gives the transient response. It is clear that if the damping, ζ, for a particular mode is positive then the transients decay in time. Conversely, for negative damping, disturbances grow and the system is unstable. Instability is a major consideration in rotating machinery, fundamentally because energy in continually supplied to the rotor by virtue of its rotation.

3.5 Further Modelling Considerations

Much of the power of rundown curves derives from the fact that they can easily relate to predictions from theoretical (almost always FE) models: it is often the comparison of models and plant data which really contributes to a thorough understanding of plant behaviour. For this reason, a very brief overview of FE methods is given in Section 3.2, but for an in-depth discussion, the reader is directed towards more detailed texts (see, e.g. Friswell et al. 2010). In this section, some further analysis is shown which may be used to enhance the usefulness of numerical analyses.

3.5.1 Mode Shapes

The general equation motion is given in Equation 3.24, and it is clear that the motion can be described with matrices to represent the mass stiffness and gyroscopic terms of the system. There are several ways in which these matrices are obtained, the most common being the FEM. It is clear from the basic theory of linear equations that, under some circumstances, non-zero solutions may be obtained even when there is no force acting on the system, and this occurs at a natural frequency of the system. With no forces applied, the equation of motion becomes (neglecting the gyroscopic term in this instance)

$$\mathbf{K}x + \mathbf{C}\dot{x} + \mathbf{M}\ddot{x} = 0 \tag{3.51}$$

Making the substitution $x = x_0 e^{j\omega t}$ we obtain the equation

$$(\mathbf{K} + j\omega\mathbf{C} - \omega^2\mathbf{M})x_0 = 0 \tag{3.52}$$

For the present, consider the case with zero damping. This equation can have a non-zero solution provided that

$$\left|\mathbf{K} - \omega^2\mathbf{M}\right| = 0 \tag{3.53}$$

This condition occurs at each of the n natural frequencies, and each of these frequencies has a corresponding displacement function or mode shape u_n which satisfies the equation

$$\left(\mathbf{K} - \omega_n^2\mathbf{M}\right)u_n = 0 \tag{3.54}$$

These mode shapes have some interesting and extremely useful properties which are now outlined. Denoting $\lambda_n = \omega_n^2$, then considering any two mode shapes n and m associated with two different natural frequencies

$$\mathbf{K}u_n = \lambda_n \mathbf{M}u_n$$
$$\mathbf{K}u_m = \lambda_m \mathbf{M}u_m \tag{3.55}$$

Multiply the first of these equation by u_m^T, the second one by u_n^T, the result is

$$u_m^T \mathbf{K}u_n - u_n^T \mathbf{K}u_m = \lambda_m u_n^T \mathbf{M}u_m - \lambda_n u_m^T \mathbf{M}u \tag{3.56}$$

It is a straightforward exercise to show that the left-hand side of this equation is zero (a consequence of the stiffness matrix being symmetric about the diagonal). Similarly, the symmetry of the mass matrix implies that

$$u_n^T \mathbf{M}u_m = u_m^T \mathbf{M}u_n \tag{3.57}$$

Then from these two relationships, we reach the very significant conclusion that

$$0 = (\lambda_n - \lambda_m)u_n^T \mathbf{M}u_m \tag{3.58}$$

Why is this significant? Because the fist term on the right-hand side is clearly non-zero. The conclusion is that any two different mode shapes are orthogonal with respect to the mass matrix, and this provides important techniques for the analysis of many aspects of dynamic behaviour.

For example consider the response of a general system to a force **F**. The equation of motion is given by

$$\mathbf{Kx} + \mathbf{C\dot{x}} + \mathbf{M\ddot{x}} = \mathbf{F} \tag{3.59}$$

If the forcing function is known, this equation can be integrated in time, but for any realistic model, this is a very time-consuming process. Often with machinery, the solution is determined in the frequency domain and is expressed as

$$x(t) = \sum X(\omega)e^{j\omega t} \tag{3.60}$$

Denoting the Fourier transform of $F(t)$ as $\tilde{F}(\omega)$, X is given by

$$X(\omega) = [\mathbf{K} + j\omega\mathbf{C} - \omega^2\mathbf{M}]\tilde{F}(\omega) \tag{3.61}$$

However, even this approach can be very demanding in computing terms if the model is detailed and/or the response is required at a number of frequencies. A more convenient approach is to express the motion in terms of the mode shapes by writing

$$x(t) = \sum_{n=1}^{N} \mathbf{u}_n q_n(t) \tag{3.62}$$

Inserting this into the equation of motion gives

$$\mathbf{K}\sum \mathbf{u}_n q_n + \mathbf{C}\sum \mathbf{u}_n \dot{q}_n + \mathbf{M}\sum \mathbf{u}_n \ddot{q}_n = \mathbf{F} \tag{3.63}$$

Now pre-multiply this equation by u_m^T then sum over all values of n

$$\sum \mathbf{u}_m^T \mathbf{K} q_n \mathbf{u}_n + \sum \mathbf{u}_m^T \mathbf{C}\dot{q}_n \mathbf{u}_n + \sum \mathbf{u}_m^T \mathbf{M}\ddot{q}_n \mathbf{u}_n = \mathbf{u}_m^T \mathbf{F} \tag{3.64}$$

Consider the summations on the left-hand side of Equation 3.64. From Equation 3.55, it may be seen that in the first summation, because $\mathbf{K}u_n = \lambda_n \mathbf{M}u_n$, the only non-zero term in the first and third summations occurs when $n = m$, so Equation 3.64 can be rewritten as

$$\lambda_m a_m q_m + \sum \mathbf{u}_m^T \mathbf{C} \dot{q}_n \mathbf{u}_n + a_m \ddot{q}_m = \mathbf{u}_m^T \mathbf{F} \qquad (3.65)$$

In this equation, $a_m = u_m^T \mathbf{M} u_m$, but because mode shapes can be scaled, it is convenient to choose the scaling so that $a_m = 1$. This leaves the damping term as the remaining difficulty: a common way around this difficulty is the assumption of so-called 'classical damping', also known as 'proportional damping' (discussed in Section 3.4.3). It is assumed that the damping matrix can be expressed as

$$\mathbf{C} = \alpha \mathbf{M} + \beta \mathbf{K} \qquad (3.66)$$

With this assumption, Equation 3.65 becomes

$$\lambda_m q_m + \alpha \dot{q}_m + \lambda_m \beta \dot{q}_m + \ddot{q}_m = \mathbf{u}_m^T F \qquad (3.67)$$

In rotating machinery, excitation is often sinusoidal and in such cases one seeks a solution of the form

$$q_m = q_{m0} e^{j\omega t}$$

Inserting this into Equation 3.59 leads to

$$q_m = \frac{\mathbf{u}_m^T \mathbf{F}}{\omega_m^2 - \omega^2 + j\omega\left(\alpha + \omega_m^2 \beta\right)} \qquad (3.68)$$

This provides a very elegant approach. From Equation 3.68, we can recover the full displacement profile $x(t)$ using Equation 3.62, and this can be done over a range of frequencies with no costly matrix inversions. The one remaining difficulty to address is the validity of the classical damping assumption: in reality, there is no fundamental justification for using it, but in many instances, it gives a reasonable approximation to reality, so let us consider why this is the case. In sinusoidal excitation, damping only plays a significant role when the excitation frequency is close to resonance, and in Equation 3.66, we have two parameters and therefore, we can model the damping at two frequencies. This is sufficient to represent the main features of many machines in practice.

It is possible, and sometimes necessary, to analyse systems with a more general form of damping. Friswell and Lees (1998) have shown how it is

FIGURE 3.9
A rotor with imbalance.

possible to express the general displacement in terms of the damped mode shapes which are, in general, complex (corresponding to a travelling elastic wave, rather than one which is standing). However, the mathematics involved is rather more cumbersome, and hence wherever possible, classical damping is assumed.

With the assumption that the damping can be regarded as proportional, the mode shapes define a natural set of coordinates which can be used to decouple the equations of motion. Whilst this is of importance in simplifying the mathematical treatment, it is even more important in physical understanding which it imparts. To illustrate this, we take an example of an unbalanced rotor. Imbalance is discussed in greater detail in Chapter 4. Consider the rotor shown in Figure 3.9 comprising a uniform shaft which carries two discs at the two one-third span positions. The two discs have equal imbalances which are opposite in phase.

In this particular example, the critical speeds (natural frequencies) occur at 15 and 40 Hz and since the rotor is symmetric, we know that the (un-normalised) mode shapes are

$$\mathbf{u}_1 = \begin{Bmatrix} 1 \\ 1 \end{Bmatrix} \qquad \mathbf{u}_2 = \begin{Bmatrix} 1 \\ -1 \end{Bmatrix} \tag{3.69}$$

It is worth noting that in this simple case, we can develop the mode shapes using physical insight. In real machines, detailed calculations will often be required, but not always. Note also that significant insight may be gleaned by considering the system's symmetry.

The damping in this example has been chosen as 2% critical in each of the modes and the forcing is proportional to the square of the rotor speed, as would be the case for imbalance excitation. Figure 3.10 shows response curve for two different excitations: in both cases, the force at the two discs is equal, but in the first case the two are in phase, whereas in the second case they are out of phase. Mathematically, we may express this by

$$\mathbf{F}_1 = a \begin{Bmatrix} 1 \\ 1 \end{Bmatrix} \qquad\qquad \mathbf{F}_2 = a \begin{Bmatrix} 1 \\ -1 \end{Bmatrix} \tag{3.70}$$

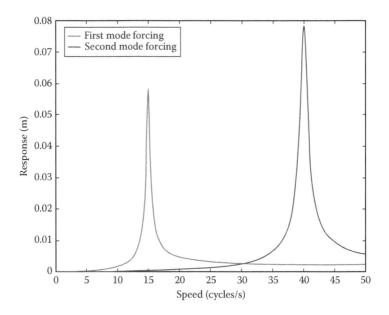

FIGURE 3.10
Imbalance response.

The important point to note here is that the first force gives no contribution to the response at the second natural frequency whilst with the second force pattern the response at the first natural frequency is unaffected by the first mode. In other words, with the second force pattern, the system does not respond at the first natural frequency. This is important in the diagnosis of several faults and is discussed in more detail in Chapter 4. This phenomena occurs because

$$\mathbf{u}_2^T\mathbf{F}_1 = a\{1 \quad -1\}\begin{Bmatrix}1\\1\end{Bmatrix} = 0 = \mathbf{u}_1^T\mathbf{F}_2 = a\{1 \quad 1\}\begin{Bmatrix}1\\-1\end{Bmatrix} \tag{3.71}$$

This is an example of how knowledge of the modes leads to a greater understanding of the dynamic behaviour of a machine or indeed any vibrating system. It is a consequence of the orthogonal properties of the normal modes.

3.5.2 Perturbation Techniques

It is often very useful to gain some insight into the sensitivity of a model to small changes.

Provided that the change in stiffness or damping term is not too great, in a sense that will be discussed later, it is reasonable to assume that the motion can be described in terms of the un-damped natural frequencies and mode shapes. This bypasses the need for the analysis of the problem for each change which may be considered.

To develop this approach, we consider a general structure neglecting damping. Using the FE study of the preceding section, the equation of motion may be written as

$$[\mathbf{K} + \Delta\mathbf{K}]\{y\} - \omega^2[\mathbf{M} + \Delta\mathbf{M}]\{y\} = 0 \tag{3.72}$$

We now consider the effect of 'turning on' the perturbing effect using a parameter denoted by λ. The deflection of the nth mode may be written as

$$\{y_n\} = \{y_{0n}\} + \lambda\{y_{1n}\} + \lambda^2\{y_{2n}\} + \cdots \tag{3.73}$$

whilst the corresponding natural frequency is written as

$$\omega_n = \omega_{0n} + \lambda\omega_{1n} + \lambda^2\omega_{2n} + \cdots \tag{3.74}$$

These expansions are now inserted into Equation 3.72 giving

$$[\mathbf{K} + \lambda\Delta\mathbf{K}]\{y_{0n} + \lambda y_{1n} + \lambda^2 y_{2n} + \cdots\}$$

$$- (\omega_{n0} + \lambda\omega_{n1} + \lambda^2\omega_{n2})^2 [\mathbf{M} + \lambda\Delta\mathbf{M}]\{y_{0n} + \lambda y_{1n} + \lambda^2 y_{2n}\} = 0 \tag{3.75}$$

From this, an equation for each power of λ is given, hence for λ^0:

$$[\mathbf{K}]\{y_{0n}\} - \omega_{0n}^2[\mathbf{M}]\{y_{0n}\} = \{0\} \tag{3.76}$$

for λ^1

$$\mathbf{K}y_{1n} + [\Delta\mathbf{K}]y_{0n} - \omega_{0n}^2\mathbf{M}y_{1n} - \Delta\mathbf{M}\omega_{0n}^2 y_{0n} - 2\omega_{0n}\omega_{1n}\mathbf{M}y_{0n} = \{0\} \tag{3.77}$$

Equations can be written for higher powers of λ and these become progressively more complicated. However, for the present study, we only require the zeroth and first power of lambda.

Equation 3.76 is simply a restatement of the problem with no perturbation, and this is conveniently solved using standard methods: hence, the frequencies ω_{0n} and the corresponding mode shapes $\{y_{0n}\}$ are known and furthermore, these mode shapes form a complete orthonormal set. The first-order changes $\{y_{1n}\}$ are expressed in term of these known modes

$$y_{1n} = \sum a_{nm} y_{0m} \tag{3.78}$$

Inserting this expression in Equation 3.77 gives

$$\left[\mathbf{K} - \omega_{0n}^2\mathbf{M}\right]\left\{\sum a_{nm}y_{0m}\right\} + \left[\Delta\mathbf{K} - \omega_{n0}^2\Delta\mathbf{M}\right]y_{0n} - 2\omega_{0n}\omega_{1n}[\mathbf{M}]y_{0n} = \{0\} \qquad (3.79)$$

Now taking $n = m$, the first on the left-hand side of Equation 3.79 becomes zero. The equation is now pre-multiplied by $\{y_{0n}\}^T$, and using the orthogonality relationships, we obtain

$$y_{0n}{}^T\left[\Delta K - \omega_{n0}^2\Delta\mathbf{M}\right]y_{0n} - 2\omega_{0n}\omega_{1n} = \{0\} \qquad (3.80)$$

This gives the first-order frequency changes as

$$\omega_{1n} = \frac{y_{0n}{}^T\left[\Delta\mathbf{K} - \omega_{0n}^2\Delta\mathbf{M}\right]y_{0n}}{2\omega_{0n}} \qquad (3.81)$$

It is straightforward to insert this expression back into Equation 3.79 and assume $n \neq m$ and deduce and equation for a_{nm} giving the changes in mode shape. The form of this expression is physically clearer by noting that the changes in frequency squared may be written as

$$\Delta\omega_n^2 = 2\omega_{0n}\omega_{1n} = y_{0n}{}^T\left[\Delta\mathbf{K} - \omega_{0n}^2\Delta\mathbf{M}\right]y_{0n} \qquad (3.82)$$

Before continuing the calculation, note that a reconsideration of Equation 3.79 for the case $n \neq m$ leads to

$$a_{nm} = \frac{\{y_{0n}\}^T[i\omega_{0n}\mathbf{C}]\{y_{0m}\}}{\omega_{0n}^2 - \omega_{0m}^2} \qquad (3.83)$$

Equations 3.81 and 3.83 give the first-order perturbation solutions to the problem of a general system. The analysis may be extended to examine the accuracy of the calculation, by assessing the second-order contributions, but this is beyond the scope of the present discussion. The perturbation approach is discussed by Wilkinson (1965), Fox and Kapoor (1968). A specific application is given by Lees (1999).

Example 3.2

Calculate the natural frequency of a uniform beam, then with a mass of 2 t at one quarter span. Let's take the span between supports as 5 m and the diameter as 0.5 m. Note that for consideration of the first mode, Euler–Bernoulli theory will suffice.

As mass is added to the rotor, its natural frequency will decrease. Considering only the variation in the frequency of the lowest mode, we denote the amplitude of the (mass normalised) mode shape at the location where the mass is added as $y_{1L/4}$, then for any mass addition the change in natural frequency is given (from Equation 3.81) as

$$\Delta\omega_1 = -\omega_1\Delta\mathbf{M}y_{1L/4}^2$$

This equation is, of course, valid only for the additions of small masses. As the changes become larger, the mode shape begins to alter as shown in Equation 3.68, and the calculation becomes somewhat more complex. Nevertheless, this approach is helpful in two ways; it is helpful in making quick assessments, and perhaps more importantly, it gives insight into parameters of significance.

Figure 3.11 shows the variation of the lowest natural frequency with the addition of a range of masses up to 2500 kg. The frequency was recalculated fully for each case, but the graph also showed the prediction of perturbation theory (to first order).

It is clear that the perturbation prediction provides a good estimate of the frequency changes, although there is a gradual deterioration in its accuracy as the magnitude of the additional mass increases. But more information can be inferred directly. For example a direct assessment can be made about the change in natural frequency if the mass is added at the central node rather than at $L/4$. The appropriate variation can be

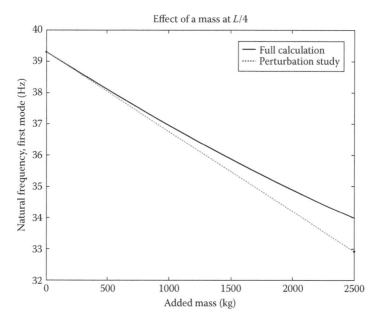

FIGURE 3.11
Comparison of 'exact' and perturbation solutions.

obtained by considering the relative magnitudes of the mode shapes at quarter and mid-span, respectively:

$$y_{1L/4} = 0.0114$$

$$y_{1L/2} = 0.0161$$

This means that (from Equation 3.81) the sensitivity of the system to a mass at mid-span is r times greater than that at quarter span, where r is given by

$$r = \left(\frac{y_{1L/2}}{y_{1L/4}}\right)^2 = 1.987$$

Without further calculation, we can say that a given mass will have approximately double the effect when placed at the mid-span rather than the quarter span position. It is precisely this facility to visualise the effects of change without direct calculation which makes this technique so useful. The explicit calculation of the influence at mid-span is shown in Figure 3.12 together with the scale perturbation line.

It should be noted that the application of this techniques enables the analyst to significantly enhance the understanding gained from a single numerical calculation.

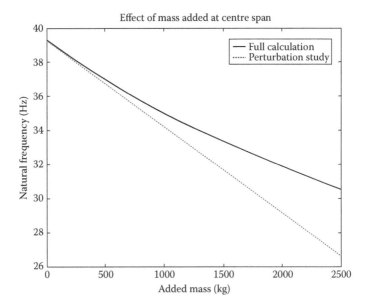

FIGURE 3.12
Comparison of exact and perturbed solutions.

3.6 Summary

Data from rotating machinery are rich in information but often complex and dependent on many parameters. The effective diagnosis of plant issues demands the effective assimilation of data, and the various ways in which data can be presented play an important part in forming an overall judgement on the internal processes within a machine. This chapter presents an overview of the various ways of viewing data, together with some analysis and modelling approaches. Almost invariably some combination of these approaches is required to deal with real plant problems. Having set out basic methods, Chapters 4, 5 and 6 discuss some of the more common type of problems arising on rotating machines.

Problems

3.1 A uniform steel shaft 1.5 m in length is supported in short rigid bearings at each end. The shaft has a diameter of 20 mm. A series of steel discs, each 1 cm thick is available for mounting at the mid-span location. Calculate the natural frequency with discs of diameter 50, 100 and 150 mm.

3.2 A shaft 1 m in length is supported at each end, on two identical bearings with a constant stiffness of 10^6 N/m. The diameter is 20 mm and it has two discs mounted 0.25 m from each bearing. Both discs have diameter 100 mm and thickness 10 mm. Using four Euler FEs, calculate the first two natural frequencies. What happens if the number of elements is doubled?

3.3 Repeat Problem 2 using Timoshenko beam elements (thereby including shear and rotary inertia effects). Comment on your answers.

3.4 Using the analytic expression for the natural frequency of a pinned–pinned Euler beam and observing that the mode shapes are sinusoidal, calculate the first natural frequency when each of the three discs of problem 1 is applied at the centre span.

3.5 A large fan is mounted on two short rigid bearings 2 m apart. The shaft is made of steel 0.2 m in diameter and the fan impeller is overhung by 0.5 m. The impeller has a mass of 800 kg and can be regarded as a disc of diameter 1.5 m. Draw the Campbell diagram and determine the shaft speeds at which speed equals natural frequency. These are the critical speeds.

References

Fox, R.L. and Kapoor, M.L., 1968, Rates of change of eigenvalues and eigenvectors, *American Institute of Aeronautics and Astronautics Journal*, 6(12), 2426–2429.

Friswell, M.I. and Lees, A.W., 1998, Resonance frequencies of viscously damped structures, *Journal of Sound & Vibration*, 217(5), 950–959.

Friswell, M.I., Penny, J.E.T., Garvey, S.D. and Lees, A.W., 2010, *Dynamics of Rotating Machines*, Cambridge University Press, New York.

Lees, A.W., 1999, The use of perturbation theory for complex modes, in *17th International Modal Analysis Conference*, Kissimmee, FL.

Wilkinson, J.H., 1965, *The Algebraic Eigenvalue Problem*, Clarendon Press, Oxford, U.K.

Zienkievicz, O.C., Taylor, R.L., and Zhu, J.Z., 2005, *The Finite Element Method*, 6th edn., Elsevier Butterworth-Hienemann, Oxford, U.K.

4

Faults in Machines – Part 1

4.1 Introduction

The purpose of all monitoring is the understanding of the current state of the plant. In order to place in context the work presented in other chapters, it is essential to discuss the manifestation of various types of plant fault. The way in which a given machine will respond to an error type is, of course, partly dependent on the geometry of the machine in question: nevertheless, several of the more important categories of fault may be described generically. Before embarking on a discussion of the more esoteric faults, it is important to examine the properties of rotor unbalance. In Chapter 3, imbalance was briefly described and used as an example to illustrate several techniques. As imbalance is the most prevalent of all rotor faults, further details are given of approaches to the identification and rectification of imbalance, but an important consideration in this is the flexibility of the rotor itself. We begin, therefore, with a discussion on the difference between rigid and flexible rotors as these issues have an important influence on the optimum approaches to rectify problems.

4.2 Definitions: Rigid and Flexible Rotors

It is important to study machine imbalance for two reasons: First, although it is now well understood, it remains the most common single cause of high vibration in rotating machinery. Second, the way in which local imbalances transform to give distributed vibration responses helps to gain an insight into the techniques at our disposal for the discussion of other phenomena.

The discussion here will be for a general flexible rotor – but we immediately run into a difficulty of terminology. In some older textbooks and papers, a rotor was said to be flexible if it operated above 70% of its first critical speed. For our purposes, this is a very inadequate definition. Reiger (1986) gives a far more meaningful definition: 'A flexible rotor is defined as being any rotor

FIGURE 4.1
A rigid rotor on spring supports.

that cannot be effectively balanced throughout its speed range by placing suitable correction weights (sic) in separate planes along its length'. To clarify we must distinguish between the rotor and the machine. The rotor is, self-evidently, just that component which rotates. So with this distinction let us consider the simple machine shown in Figure 4.1. The mass of the rotor is 100 kg and each bearing has a stiffness of 5×10^6 N/m. This gives a resonant frequency of 50.3 Hz. Given that the rotor rotates at 2400 rev/min, the speed represents 80% of the natural frequency and so the machine may be called flexible. It will respond significantly to imbalance as we will discuss shortly.

However, the rotor is rigid. It moves simply as a rigid body having four degrees of freedom, two translational and two rotational. The shape of the first mode is shown as a dotted outline in Figure 4.1; it is simply a uniform translation from the equilibrium position. This mode is sometimes referred to as the 'bounce' mode and can, of course, arise in the two directions orthogonal to the axis of the rotor. Note that motion in the axial direction is not excited by unbalance and torsional motion has been neglected for this case. In fact, of course, there is no such thing as a truly rigid rotor; nevertheless, the concept is convenient and for a large number of machines it represents a convenient abstraction and a reasonable approximation. Friswell et al. (2010) carry a thorough discussion of the dynamics of a rigid rotor, but the requirement here is to ascertain the validity of the concept for a given case.

Restricting attention to the plane of the page, it is clear that the rotor can only move in two ways (or combinations thereof): It can either move with equal amplitude at all points along the length or rotate about the central point. Any other possible motion is simply a (linear) combination of these two basic modes of vibration. Figure 4.2 shows the second mode which exists for a rigid rotor in each direction. This is often called the 'rocking' mode – again it occurs in both directions.

In general, of course, the rotor will move in the two orthogonal directions, and hence, there will be a total of four degrees of freedom. Perhaps the most obvious way of expressing the four degrees of freedom is in terms of the displacements at the two bearings, but this choice is not unique: one could, for instance, choose the coordinates at any two points along the rotor. A less obvious notation leads to a rather more compact description of the mechanics of the system. Following Friswell et al. (2010), the motion of the rotor is

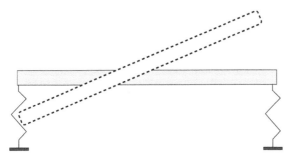

FIGURE 4.2
Second mode ('rocking').

described in terms of the motion of the centre of gravity of the rotor and its rotations about the x and y axes. The equations of motion will be given by

$$k_{x1}(u - a\psi) + k_{x2}(u + b\psi) + m\ddot{u} = 0$$

$$k_{y1}(v + a\theta) + k_{y2}(v - b\theta) + m\ddot{v} = 0$$

$$+ ak_{y1}(v + a\theta) - bk_{y2}(v - b\theta) + I_p\Omega\dot{\psi} + I_d\ddot{\theta} = 0$$

$$- ak_{x1}(u - a\psi) + bk_{x2}(u + b\psi) - I_p\Omega\dot{\theta} + I_d\ddot{\psi} = 0$$

(4.1)

where
   u, v represent the motions of the centre of gravity in the x and y directions
   θ, ψ are the rotations about the x and y axes, respectively
   I_d and I_p are the diametral and polar moments of inertia respectively
   Ω is shaft speed

The centre of mass is a distance a from bearing 1 and b from bearing 2. These equations are formulated simply by expressing Newton's law for both the translation and the rotations. The two terms involving $I_p\Omega$ are the gyroscopic moments which arise basically from the conservation of angular momentum. A full discussion of these moments are given by Friswell et al. (2010), but for many rotors, particularly in arrangements with bearing outboard of the main inertia components, the influence of these terms will be minor. They are included here for the sake of completeness as they become important in machines with overhung rotors or those with very large impellers. Simply rearranging these equations yields

$$(k_{x1} + k_{x2})u + (-ak_{x1} + bk_{x2})\psi + m\ddot{u} = 0$$

$$(k_{y1} + k_{y2})v + (ak_{y1} - bk_{y2})\theta + m\ddot{v} = 0$$

$$(ak_{y1} - bk_{y2})v + \left(a^2 k_{y1} + b^2 k_{y2}\right)\theta + I_p\Omega\dot{\psi} + I_d\ddot{\theta} = 0$$

$$(-ak_{x1} + bk_{x2})u + \left(a^2 k_{x1} + b^2 k_{x2}\right)\psi - I_p\Omega\dot{\theta} + I_d\ddot{\psi} = 0$$

(4.2)

Letting

$$k_{xT} = k_{x1} + k_{x2} \quad k_{yT} = k_{y1} + k_{y2}$$

$$k_{xC} = -ak_{x1} + bk_{x2} \quad k_{yC} = -ak_{y1} + bk_{y2} \tag{4.3}$$

$$k_{xR} = a^2 k_{x1} + b^2 k_{x2} \quad k_{yR} = a^2 k_{y1} + b^2 k_{y2}$$

then (4.2) can be written more concisely thus

$$k_{sT} u + k_{xC} \psi + m\ddot{u} = 0$$

$$k_{yT} v - k_{yC} \theta + m\ddot{v} = 0$$

$$-k_{yC} v + k_{yR} \theta + I_p \Omega \dot{\psi} + I_d \ddot{\theta} = 0 \tag{4.4}$$

$$k_{xC} u + k_{xR} \psi - I_p \Omega \dot{\theta} + I_d \ddot{\psi} = 0$$

Of course because of the gyroscopic terms, if the motion in the u and v directions is coupled to some extent, and so, in general, the four equations must be solved together to reflect the behaviour of the rotor system. It is important to appreciate that here, the rotor is rigid, and consequently, the motion of the two ends (or specifically, bearing locations) is determined, and the dynamics of the system is completely specified. Denoting the two bearing locations as A and B and their respective motions as $x_A(t)$, $y_A(t)$, $x_B(t)$, $y_B(t)$, the motion of the centre of mass (gravity) may be expressed as

$$x_C = \frac{ax_A + bx_B}{a+b} \qquad y_C = \frac{ay_A + by_B}{a+b} \tag{4.5}$$

This is where we encounter a major distinction because this is not the case if the rotor is flexible. Flexible rotors form a major part of large machines in particular, but there is a long-term trend towards their use in faster machines. Where the rotor is flexible, the relationship of the points A, B and C is more complicated and is inherently frequency dependent.

It is helpful to define the following parameters:

$$f_x = x_C - \frac{x_A + x_B}{2} \quad f_y = y_C - \frac{y_A + y_B}{2} \tag{4.6}$$

and with these defined, a measure of the degree of flexibility in the rotor is just

$$r_f = \sqrt{f_x^2 + f_y^2} \tag{4.7}$$

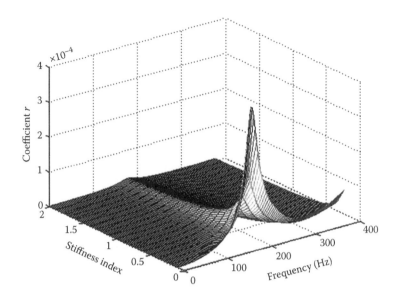

FIGURE 4.3
Variation of modes with rotor stiffness.

By measuring the location of the shaft at any three axial positions, a measure of this type can be defined. Figure 4.3 shows a plot of how the parameter r_f varies as the flexibility of the rotor is changed. This calculation used a uniform beam divided into four elements, and flexibility was altered by varying the value of Young's modulus by the stiffness factor.

In this calculation, stiffness was varied by adjusting Young's modulus. However, with less than three locations measured, no definite conclusions can be made concerning the shaft dynamics without some further assumptions or, best of all, a reliable theoretical model of the machine. Of course, this is no easy task, and other chapters in this book are devoted to the derivation of such models. As mentioned elsewhere (see Chapter 7), the modelling of the rotor itself is the easiest task in the representation of a complete machine. We, therefore, examine the properties of the rotor in terms of its own dynamics.

The simplest case we can consider is a truly rigid rotor with a mass M and a diametral moment of inertia about its centre of gravity of I_d. It is easily shown that the two natural frequencies for a rigid rotor (neglecting gyroscopic terms) are given by

$$\omega_1 = \sqrt{\frac{k_T}{m}} \qquad \omega_2 = \sqrt{\frac{k_R}{I_d}} \qquad (4.8)$$

Consider the following example. A uniform rotor is mounted on two idealised bearings: as an exercise the dimensions and mass of the rotor are fixed, but Young's modulus is varied having the effect of changing the rotor's free-free natural frequency, or to be more specific, its first bending (free-free) natural frequency. Note that for a real rotor this mode will be the seventh free-free mode as there will be six rigid bode modes (with zero frequency) corresponding to translation along each of the three coordinates and rotations about each of the three axes. The frequency of this first flexural free-free mode provides a very effective indicator as to the behaviour of the rotor in a machine. It is easily calculated from a rotor model, and since for this evaluation one is not dependent of the uncertainties of bearings and support structures, these predictions tend to be reliable. The calculated values are readily validated by suspending the rotor in slings from a crane and then applying an impact test, preferably in the horizontal direction to minimise the effect of the suspension system.

Now consider the effect of introducing rotor flexibility; Equation 4.4 describes the motion of a rigid rotor on a support structure, but clearly if the shaft itself can flex, this decreases the stiffness of the system over and consequently lowers the first natural frequency.

Figure 4.4 shows the variation in overall natural frequency as the stiffness of the rotor itself is varied (note that the bearing stiffness is held constant). A uniform rotor has been used for these calculations, but a more general case would differ only in detail. In the graph each of the axes has

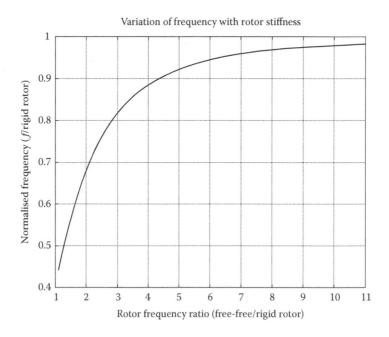

FIGURE 4.4
Influence of rotor stiffness on system resonance.

TABLE 4.1

Variation of System Natural Frequencies of 1 m Shaft

Free-Free Frequency (Hz)	Mode 1 (Hz)	Mode 2 (Hz)
67.7	27.7	88.8
117.3	42.1	106.3
151.3	48.8	110.6
179.1	52.7	112.5
203.1	55.3	113.6
224.5	57.2	114.5
244.1	58.5	114.9
262.2	59.6	115.2
279.1	60.4	115.5
∞ (rigid)	67.8	117.4

been normalised in terms of the frequency of the rigid rotor case first (symmetric) mode. The graph shows, for example, that if the rotor's first flexural free-free natural frequency is twice the first natural frequency of the rigid rotor mounted on the bearings, then the overall machine natural frequency is reduced to 68% of the rigid rotor case. If, on the other hand, the free-free frequency is ten times that of the rigid rotor machine, then the reduction is only about 3%. Apart from changes in frequencies, there are some important distinctions between the properties of rigid and flexible rotors which are discussed in the section on balancing.

It is clear from the graph, for instance, that if the free-free of the rotor is four times the rigid rotor frequency, then the effect of the rotor flexibility is to reduce the system frequency by about 12%.

An example of this is illustrated in Table 4.1. A uniform rotor is 1 m between simple bearings, each with a stiffness of 5×10^5 N/m, the diameter is 3 cm, the density is that of steel, but Young's modulus has been varied over a range in order to change the rigidity.

Using the same model it is helpful to examine the mode shapes as well as the natural frequencies. The most salient question is the location and magnitude of the maximum displacement (or, quite possibly, maximum stress). With a rigid rotor (corresponding to an infinite bending free-free frequency), the amplitude of vibration in the first mode is clearly equal to the amplitude at bearing 1 – indeed it is uniform. However, as the free-free natural frequency reduces, shaft bending changes this situation and the peak amplitude may be substantially different from that measured at the bearing and the location of this peak amplitude may well be elsewhere along the length of the rotor. Figure 4.5 shows this variation.

Hence, it is seen that for this case of a simple beam if the free-free natural frequency of the shaft is double the rigid rotor resonance, then the maximum

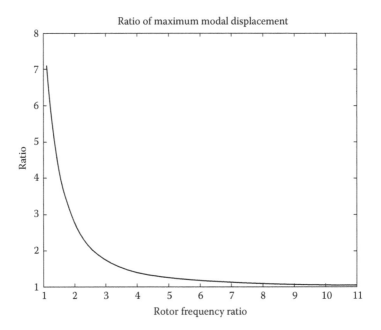

FIGURE 4.5
Modal displacement ratios.

amplitude is about 2.5 times the amplitude at the bearing location. Of course in a real situation, the most appropriate means of evaluating these concepts is by the use of an appropriate model: the simple case here serves to illustrate the point.

There is another very important aspect of the distinction between rigid and flexible rotors which is closely related to this discussion: a rigid rotor has precisely six free-free modes relating to the six degrees of freedom. These modes correspond to the uniform motion along the axes (two transverse modes in each direction, one torsional and one extentional). A flexible rotor by contrast has many more free-free modes. This means that the motion of a flexible rotor cannot be described in terms of its six rigid body modes, a point which has important consequences in rotor balancing, as discussed in Section 4.2.

Many machines are designed to operate above their second critical speed, and in these circumstances, it may be unclear whether the rotor is operationally rigid or flexible. Nevertheless, it is important to draw the distinction as this will dictate the form of balancing needed; are two planes sufficient or are more required? Commercial considerations dictate the absolute need to do no more than necessary. This is an important example of the many instances in which the operator must have some understanding of the machines under his or her control.

4.3 Mass Imbalance

4.3.1 General Observations

The response of rotors to mass imbalance is without doubt the most widely encountered single problem in rotating machinery. Despite being well understood for many years, it is still prevalent in a wide range of turbo-machines. The difficulties arise from the strong sensitivity of high-speed machines to relatively small manufacturing tolerances. If there are no external forces acting on the system, then the equation of motion for free vibration is given by

$$\mathbf{Kx} + \mathbf{C\dot{x}} + \mathbf{M\ddot{x}} = 0 \qquad (4.9)$$

But to understand the implications of this equation when applied to a rotating machine, careful consideration must be given to the individual terms. In the inertial term, the relevant acceleration is that of the centre of mass of the rotor, rather than its geometric centre, the two differing by the mass eccentricity. The issue here is that because the centre of mass does not coincide with the geometric centre (or, more importantly, the centre of rotation), there is an acceleration imposed on the centre of mass by the rotation of the shaft. The difference in location may be expressed as

$$\mathbf{x}_G = \left\{ \begin{matrix} \mathbf{u}_G \\ \mathbf{v}_G \end{matrix} \right\} = \left\{ \begin{matrix} \mathbf{u} + \varepsilon \cos\phi \\ \mathbf{v} + \varepsilon \sin\phi \end{matrix} \right\} \qquad (4.10)$$

The eccentricity ε will usually be a vector giving the mass eccentricity at each point along the rotor.

In inertial term of Equation 4.9, $\mathbf{x}$ must be replaced with $\mathbf{x}_G$, so that the equation becomes

$$\mathbf{Kx} + \mathbf{C\dot{x}} + \mathbf{M\ddot{x}}_G = 0 \qquad (4.11)$$

For constant speed of Ω radians per second, this leads to

$$\left(\mathbf{K} + j\mathbf{C}\Omega - \mathbf{M}\Omega^2 \right)\mathbf{x} = \mathbf{M}\Omega^2 \varepsilon e^{j\Omega t} \qquad (4.12)$$

The steady-state solution to this equation must be of the form $\mathbf{x} = \mathbf{x}_0 e^{j\Omega t}$ and substitution of this form into Equation 4.12 enables cancellation of the sinusoidal variation to give

$$\left(\mathbf{K} + j\mathbf{C}\Omega - \mathbf{M}\Omega^2 \right)\mathbf{x}_0 = \mathbf{M}\Omega^2 \varepsilon \qquad (4.13)$$

A direct solution can be expressed as

$$\mathbf{x}_0 = [\mathbf{K} + j\mathbf{C}\Omega - \mathbf{M}\Omega^2]^{-1}\mathbf{M}\Omega^2\varepsilon \qquad (4.14)$$

Without discussing any details of the solution to the motion, three features are immediately obvious:

a. The response is synchronous, that is in steady-state operation, the vibration is at rotor speed.

b. As the speed tends to zero, the response also tends to zero.

c. As the speed increases, the vibration tends to a finite limit.

The third of these observations is perhaps worthy of note as it has a very simple physical interpretation: from Equation 4.14, it is clear that as $\Omega \to \infty$, then $y_0 = -\varepsilon$. This is simply saying that as the speed increases, the rotor will tend to rotate about its centre of mass rather than the geometric centre, which of course will be observed as a vibration. Indeed, on reflection this is an inevitable result as any other scenario would give rise to infinite forces.

It is instructive to extend this examination of mass imbalance by considering the modal nature of the solutions. Whilst this is very useful in some numerical studies significantly reducing computing time, the real benefit lies in the insight it offers in understanding the nature of the phenomena.

Considerable insight into imbalance response can be gained by using some properties of eigenvalues and eigenfunctions (corresponding to natural frequencies and mode shapes). Although this requires the use of some algebra, the overall simplification is very considerable.

Provided that the damping may be regarded as 'classical' or 'proportional', the modes of this system form a complete orthonormal set and so the unknown response may be written as

$$x_0(z) = \sum_{n=1}^{\infty} a_n \psi_n(z) \qquad (4.15)$$

Note that the so-called 'proportional' model assumes that the damping matrix is a linear combination of the mass and stiffness matrices. As discussed in Chapter 3, there is no physical basis for this assumption, but it significantly simplifies the calculation without, in many cases, significantly altering the model predictions. In fact this assumption on damping is rarely valid: in many rotating machines, the damping will be concentrated at the bearings. However, for cases with modest damping, the assumption leads to enhanced understanding.

This is now inserted into Equation 4.13, and the resulting equation is pre-multiplied by ψ_m^T. This gives

$$\sum_{n=1}^{\infty} a_n \psi_m^T \mathbf{K} \psi_n + j\Omega \sum_{n=1}^{\infty} a_n \psi_m^T \mathbf{C} \psi_n - \Omega^2 \sum_{n=1}^{\infty} a_n \psi_m^T \mathbf{M} \psi_n = \psi_m^T \mathbf{M}\varepsilon \qquad (4.16)$$

From the orthonormality conditions (the basic properties of the mode shapes)

$$\psi_m^T \mathbf{K} \psi_n = \omega_n^{\,2} \delta_{nm}$$

$$\psi_m^T \mathbf{C} \psi_n = \left(\alpha + \beta \omega_n^{\,2} \right) \delta_{nm} \qquad (4.17)$$

$$\psi_m^T \mathbf{M} \psi_n = \delta_{nm}$$

where δ_{nm} is the Kronecker delta. Using these relationships in Equation 4.16, we see that for each value of m, there is only a contribution from $n = m$; then

$$a_m \left(\omega_m^2 + j\Omega \left[\alpha + \beta \omega_m^2 \right] - \Omega^2 \right) = \psi_m^T \mathbf{M} \Omega^2 \varepsilon \qquad (4.18)$$

and so

$$a_m = \frac{\psi_m^T \mathbf{M} \Omega^2 \varepsilon}{\left(\omega_m^2 + j\Omega \left[\alpha + \beta \omega_m^2 \right] - \Omega^2 \right)} \qquad (4.19)$$

Hence, the full response can be evaluated. Rotor unbalance is an ever-present feature of rotating machinery, and the resulting synchronous vibration is characteristic of, but not uniquely related to, unbalance. As discussed in Section 4.4, a rotor bend also generates a synchronous vibration signal, yet some aspects of its behaviour are in marked contrast to pure imbalance.

Before proceeding it is worth reflecting on Equation 4.19. In essence it is a rather simple expression for the modal content of the response. The essential message is that the excitation component in the mth mode is proportional to $\psi_m^T \mathbf{M}\varepsilon$ multiplied by a fairly complicated function of shaft speed. This carries some important insights: for instance, in the case of a uniform rotor, an imbalance at the centre span will have no influence on the second mode because the second mode has zero amplitude at this point along the rotor. Indeed, a comparison of the relative amplitudes at different rotor speeds can often be used to locate (axially) imbalance or other faults. The concept involved is now illustrated with an example. A detailed model of a machine can offer tremendous insight into operational issues, but even limited knowledge can be used to advantage, albeit with some level of uncertainty.

Example 4.1

To illustrate this, consider the rundown profile shown in Figure 4.6. The curve shows two distinct peaks at 480 and 1900 rev/min. If it is postulated that the imbalance is dominated by a single mass, what can be said about the axial location of this imbalance?

Provided there is knowledge, even if approximate, of the mode shapes associated with the two critical speeds, then some assessment can be made. In this case we assume that the mode shapes are those of a uniform rotor, illustrated in Figure 4.7.

From Figure 4.6 it is clear that the first peak is about five times higher than the second peak. The presence of a peak at 1900 rev/min immediately indicates that the imbalance (if there is a single imbalance) cannot be at the mid-span. The damping of the two modes (as assessed by the ratio of the peak width to the resonant frequency) is rather different, and from the figure it appears that the damping of mode two is about three times that of the first mode. On this assumption we may write

$$x(\omega_1) = a \sin\left(\frac{\pi z_u}{L}\right) \quad x(\omega_2) = a \sin\left(\frac{2\pi z_u}{L}\right)$$

So that $\dfrac{x(\omega_1)}{x(\omega_2)} = 5 = \dfrac{3a \sin\left(\dfrac{\pi z_u}{L}\right)}{a \sin\left(\dfrac{2\pi z_u}{L}\right)}$

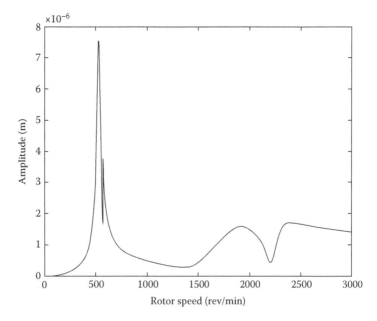

FIGURE 4.6
Transient response curve.

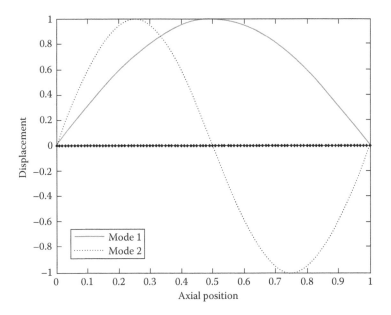

FIGURE 4.7
Mode shape estimates.

$$\text{Therefore, } 10 \times \sin\left(\frac{\pi z_u}{L}\right)\cos\left(\frac{\pi z_u}{L}\right) = 3 \times \sin\left(\frac{\pi z_u}{L}\right)$$

Giving $z_u = \cos^{-1}(0.3) \times L/\pi \approx 0.4\,L$

There is, of course, some uncertainty in the absence of some damping model, but reasonable estimates can be established using knowledge of the mode shapes.

4.3.2 Rotor Balancing

To overcome, or at least limit, the influence of unbalance, a range of balancing techniques have been developed. Early work was based on rigid rotors, but in more recent years developments have focused on flexible rotors as these have assumed ever greater importance. The field was reviewed by Parkinson (1991) who quotes some 100 references, and more recent work has been surveyed by Foiles et al. (1998) citing a further 160 papers. Although the basic methods are now reasonably well understood (at least at the research level), the large number of papers on this topic reflects the very high commercial importance. There can be little doubt that unbalance is the predominant single mechanism of concern in all rotating machinery.

The approaches to balancing may be categorised into two basic approaches: the first of these investigates the influence of added masses and a fixed speed, whilst in the second approach balancing is performed at or near to each mode in turn. Both these approaches rely on the fundamental assumption of linearity: in practice corrections are sometimes made for non-linearity by additional test runs, thereby successively seeking to reduce the unbalance. In the case of the modal approach, however, there are some rather more rigorous requirements.

In principle all rotor balancing is based on the extrapolation of the influence of a small added mass to the full negation of any initial unbalance. In its simplest form the process will be, within some tolerable limit, linear.

4.3.2.1 Influence Coefficient Approaches: Single Plane

The concept here is very straightforward. At a fixed running speed, the influence of an added mass on the vibration is measured in terms of both magnitude and phase.

Then it is a simple matter to determine the required balance mass (frequently but incorrectly referred to as balance weights), but this scaling must done in a vector sense to give the position of the mass adjustment as well as its magnitude. To efficiently balance a machine one needs some means of identifying the angular orientation of the rotor at any given time: this might comprise a painted mark on the shaft coupled to some optical probe, or some electronic encoder system. Friswell et al. (2010) show that balancing can be achieved without a phase reference, but this involves a rather more complex procedure and should be treated only as a last resort where no phase signal is possible. The more usual approach with phase measurement is now illustrated with a simple example.

It is appropriate to balance a machine in a single plane if only one resonance influences the dynamic behaviour of the machine. This will be apparent from the shape of the rundown curve which, in these cases, will have no more than one peak in the response curve. Single-plane balancing means that balance masses can be fitted in a single axial position along the shaft. The basic idea of balancing is, in fact, very straightforward: by using trial masses, the engineer establishes the sensitivity of the system to known masses, in terms of both vibration magnitude and phase. Using this sensitivity, the magnitude and position is calculated of the mass required to counteract the initial vibration.

The operation is as follows:

a. Ensure the shaft has some form of marker which can be used to define the phase. This may take the form of a physical mark on the shaft in view of an optical probe, or some electronic encoder. The important point is to ensure a phase measurement system.

b. The vibration of the machine at the test speed is measured.

c. After halting the machine, a known mass is attached to the rotor at some known angle with respect to the phase reference.

d. The machine is then up the test speed and the vibration measured again.

e. The difference in the two vibration measurement gives the rotor's sensitivity and from this the correct mass and its required position can be calculated.

f. The machine is then stopped, the appropriate correction mass attached and the machine returned to service.

To clarify the calculation involved, let us suppose the vibration is measured with amplitude r_0 at phase θ_0, so that the motion in the x direction is given by $u = r_0\cos(\Omega t + \theta_0)$.

So, expressing this in complex form

$$u = \text{Re}\left(u_0 e^{j\theta_0}\right) \tag{4.20}$$

The motion of the rotor is described by its equation of motion:

$$ku + c\dot{u} + m\ddot{u} = m\Omega^2\varepsilon\cos\Omega t \tag{4.21}$$

Expressing this in complex form and assuming the sinusoidal form, this may be written as

$$(k + jc\Omega - m\Omega^2)u = m\varepsilon\Omega^2 e^{j(\Omega t+\phi)} = D(\Omega)u_0 e^{j\Omega t} \tag{4.22}$$

u_0 is, in general, complex. At this stage the sinusoidal terms cancel to give

$$D(\Omega)u_0 = m\varepsilon\Omega^2 e^{j\phi} \tag{4.23}$$

The point to notice here is that, although Equation 4.22 specifies the problem, it is not of much help yet because all the parameters are (usually) unknown. Since the rotor speed, Ω, is constant, the unknown mass, stiffness and damping may be lumped together into a single (unknown) parameter $D(\Omega)$ describing the full dynamic behaviour of the machine at this particular rotor speed. ($m\varepsilon$), ϕ are the parameters which are implicitly to be determined.

A trial mass m_1 is now fitted at angle ϕ_1, the machine is run back up to speed, and a new vibration u_1 is measured. The equation of motion can now be written as

$$D(\Omega)u_1 = m\varepsilon\Omega^2 e^{j\phi} + m_1 r_1\Omega^2 e^{j\phi_1} \tag{4.24}$$

Subtracting Equation 4.23 from 4.24 gives the vibration level (both amplitude and phase) resulting from the added mass, and hence this yields the sensitivity of the system, resulting in the following equation:

$$D(\Omega)(u_1 - u_0) = m_1 r_1 \Omega^2 e^{j\phi_1} \tag{4.25}$$

From which we get

$$D(\Omega) = \frac{m_1 r_1 \Omega^2 e^{j\phi_1}}{(u_1 - u_0)} \tag{4.26}$$

Notice that D is now fully defined. Using Equation 4.22, the original imbalance can be calculated as

$$D(\Omega)u_0 = \frac{m_1 r_1 e^{j\phi_1} \Omega^2 u_0}{(u_1 - u_0)} = m\varepsilon \Omega^2 e^{j\phi} \tag{4.27}$$

Taking the modulus gives the magnitude of the imbalance as

$$m\varepsilon = \left| \frac{m_1 r_1 u_0}{u_1 - u_0} \right| \tag{4.28}$$

and the orientation of the imbalance is given by

$$\phi = \tan^{-1} \left(\frac{\mathrm{imag}(D(\Omega))}{\mathrm{real}(D(\Omega))} \right) \tag{4.29}$$

This gives all the information required to balance the rotor. Balance is achieved by first removing the trial mass and then attaching an imbalance of $m\varepsilon$ at an orientation directly opposite ϕ, that is at $\phi + 180°$.

This is now illustrated with an example.

Example 4.2: Single-Plane Balancing

A machine runs at a maximum speed of 1450 rev/min and its rundown curve, shown in Figure 4.6 illustrates that there is only a single resonance (critical speed) influencing the running range. This is an important consideration as it implies that a single-plane balancing procedure will suffice to run the machine satisfactorily at 1450 rev/min. At full spend the machine shows a vibration on one bearing of 25 μm at a phase of 45° after the phase reference. A trial mass of 10 g is then attached to the rotor at a radius of 100 mm and a phase of 180° (i.e. opposite the phase marker). On running the machine back up to the same speed, the measured vibration was 20 μm at 30° before the phase reference. From this information the appropriate balancing mass to completely counteract the vibration is readily established.

Expressing the initial vibration in terms of complex numbers, a complete description of the vibration can be given by a complex number involving magnitude and phase so that

$$u_0 = 25\left(\cos\frac{\pi}{4} + j\sin\frac{\pi}{4}\right) = 25\left(\frac{\sqrt{2}}{2} + j\frac{\sqrt{2}}{2}\right)$$

Whilst representation in terms of magnitude and phase is valid, the Cartesian form is more convenient for the subsequent calculation. Note that the length scale can be expressed in any consistent units and in this instance μm are very convenient. With test mass added the amended vibration is

$$u_1 = 20\left(\cos\frac{\pi}{6} - j\sin\frac{\pi}{6}\right) = 20\left(\frac{\sqrt{3}}{2} - j\frac{1}{2}\right)$$

Hence, the vibration generated by the additional mass is given by

$$\Delta u = u_1 - u_0 = 10\sqrt{3} - \frac{25}{\sqrt{2}} + j\left(-10 - \frac{25}{\sqrt{2}}\right)$$

Our objective is to establish the unbalance (me) but, of course, the terms of $D(\Omega)$ are unknown, but in this case it has been established that

$$D(\Omega)\Delta u = -0.001\Omega^2$$

This determines $D(\Omega)$ and hence the original imbalance is given by

$$me\Omega^2 = D(\Omega)u_0 = -\frac{0.001\Omega^2 u_0}{\Delta u}$$

This implies that the balance mass must be directly opposite this position. In the current case this gives

$$me = 0.001 \times \frac{25}{\sqrt{2}}(1+j) \times \frac{1}{10\sqrt{3} - \frac{25}{\sqrt{2}} + j\left(-10 - \frac{25}{\sqrt{2}}\right)}.$$

which simplifies to $m\varepsilon = (6.4683 - 6.3035j) \times 10^{-4}$. The orientation of this imbalance is given by

$$\theta = \tan^{-1}\left(\frac{-6.3035}{6.4683}\right)$$

Then it follows that this corresponds to an imbalance of abs($m\varepsilon$) = 9.03 × 10^{-4} kg m at −44°. Hence, if the balance mass is to be added at a radius of 10 cm, then a mass of 9 g should be fitted at an angle of 134° ahead of the reference mark.

Whilst at first sight this may appear to be an involved procedure, it is readily automated. Simply stated, the engineer is observing the effect of a trial mass, then scaling this in such a way as to negate the original vibration on the machine.

Where single-plane balancing suffices, this completes the procedure. However, if the rundown curve indicates the influence of more than a single mode, then balancing in two (or sometimes more) planes is required, but very similar logic guides the process. Recall that this single-plane technique has been employed for this example because the rundown curve (in Figure 4.6) showed that only a single resonance was influencing the machine within its running range. If this is not the case, a more rigorous approach must be used, and the number of balance plane must equal the number of resonances influencing the rundown curve.

4.3.2.2 Two-Plane Balancing

Balancing in two planes requires the use of two trial masses and measurement of the response at two distinct points along the rotor. The complex vibration readings **u** will now be vectors having two (or more in the case of multi-plane) components, and the influence term **D** becomes a matrix. It should be emphasised that the procedure for two planes (or indeed, multi-planes) is, in principle, entirely equivalent to the single-plane approach. Inevitably, the calculation is slightly more complicated.

The difference is that because there are now two balance planes and two degrees of freedom, the system dynamic stiffness $D(\Omega)$ becomes a 2×2 matrix. Following Equation 4.25, two equations can now be written to express the dynamic behaviour. Denoting our two trial imbalances as $m\varepsilon_{01}$, $m\varepsilon_{02}$, note that in this case, each of terms is a vector. This is to reflect the fact that each of the two trials may involve masses on each of the two balance planes:

$$D(\Omega)\begin{Bmatrix}u_{11}\\u_{12}\end{Bmatrix} = \begin{Bmatrix}m\varepsilon_{01}e^{j\phi_{01}}\\m\varepsilon_{02}e^{j\phi_{02}}\end{Bmatrix}\Omega^2 + \begin{Bmatrix}m_{11}r_{11}e^{j\phi_{11}}\\m_{12}r_{12}e^{j\phi_{12}}\end{Bmatrix}\Omega^2 \qquad (4.30)$$

and similarly

$$D(\Omega)\begin{Bmatrix} u_{21} \\ u_{22} \end{Bmatrix} = \begin{Bmatrix} m\varepsilon_{01}e^{j\phi_{01}} \\ m\varepsilon_{02}e^{j\phi_{02}} \end{Bmatrix}\Omega^2 + \begin{Bmatrix} m_{21}r_{21}e^{j\phi_{21}} \\ m_{22}r_{22}e^{j\phi_{22}} \end{Bmatrix}\Omega^2 \tag{4.31}$$

It important that the two trial cases must be linearly independent and must involve both of the balance planes. It is convenient to change notation slightly at this point and make the r terms complex, thereby absorbing the phase information into them so that

$$R_{nk} = r_{nk}e^{j\phi_{nk}} \tag{4.32}$$

Then following the same logic as in the single-plane case, using differences between each of the two trial runs and the as-found condition, Equations 4.31 and 4.32 can be combined to give

$$D(\Omega)\begin{bmatrix} \Delta u_{11} & \Delta u_{21} \\ \Delta u_{12} & \Delta u_{22} \end{bmatrix} = \begin{vmatrix} m_{11}R_{11} & m_{21}R_{21} \\ m_{12}R_{12} & m_{22}R_{22} \end{vmatrix} \tag{4.33}$$

As before $\Delta u_{kj} = u_{kj} - u_{0j}$ represents the change introduced by the corresponding trial mass set. Equation 4.33 now encapsulates the data from the 'as-found' condition and the two trial runs and from this the matrix D can be determined as

$$D(\Omega) = \begin{bmatrix} m_{11}R_{11} & m_{21}R_{21} \\ m_{12}R_{12} & m_{22}R_{22} \end{bmatrix}\begin{bmatrix} \Delta u_{11} & \Delta u_{21} \\ \Delta u_{12} & \Delta u_{22} \end{bmatrix}^{-1} \tag{4.34}$$

The imbalance on the rotor is now given by

$$\begin{Bmatrix} m\varepsilon_1 e^{j\phi_{01}} \\ m\varepsilon_2 e^{j\phi_{02}} \end{Bmatrix} = \begin{bmatrix} m_{11}R_{11} & m_{21}R_{21} \\ m_{12}R_{12} & m_{22}R_{22} \end{bmatrix}\begin{bmatrix} \Delta u_{11} & \Delta u_{21} \\ \Delta u_{12} & \Delta u_{22} \end{bmatrix}^{-1}\begin{Bmatrix} r_{01} \\ r_{02} \end{Bmatrix} \tag{4.35}$$

This is readily expressed in terms of modulus and phase as illustrated in Example 4.3. It is important to appreciate that this estimates the position of the effective centre of mass of the rotor: Note that to compensate for this, a balance mass must be added directly opposite to these positions (i.e. 180° from those calculated).

Example 4.3: A Two-Plane Balance

A rotor is known to require balancing in two planes, and vibration measurements are made in the x direction at the two bearings. The observations of the original conditions and two trial runs are shown in the table.

Table of Readings for Example 4.3

	Bearing 1 Amplitude (mm)	Bearing 1 Phase (°)	Bearing 2 Amplitude (mm)	Bearing 2 Phase (°)
Original	0.4	120	0.3	240
100 g at 150 mm 0°, disc 1	0.5	140	0.35	150
100 g at 150 mm 0°, disc 2	0.45	50	0.4	300

Following the steps from Equations 4.30 to 4.36, the parameters are as follows.

(Note that here all measurements are kept in millimetres: any consistent set of units is acceptable.)

The as-found vibration is

$$\mathbf{u}_0 = \begin{Bmatrix} u_{01} \\ u_{02} \end{Bmatrix} = \begin{Bmatrix} 0.4 \times e^{j \times 120 \times \pi/180} \\ 0.3 \times e^{j \times 240 \times \pi/180} \end{Bmatrix} = \begin{Bmatrix} -0.2 + 0.3464j \\ -0.15 - 0.2598j \end{Bmatrix} \text{mm}$$

On the first trial run

$$\mathbf{f}_1 = \begin{Bmatrix} m_{11}R_{11} \\ m_{12}R_{12} \end{Bmatrix} = \begin{Bmatrix} 0.1 \times 0.15 \\ 0 \end{Bmatrix} \text{kg m}$$

and the trial response is

$$\mathbf{u}_1 = \begin{Bmatrix} u_{11} \\ u_{12} \end{Bmatrix} = \begin{Bmatrix} 0.5 \times e^{j \times 140 \times \pi/180} \\ 0.35 \times e^{j \times 150 \times \pi/180} \end{Bmatrix} = \begin{Bmatrix} -0.383 + 0.321j \\ -0.303 - 0.175j \end{Bmatrix} \text{mm}$$

Hence

$$\Delta\mathbf{u}_1 = \mathbf{u}_1 - \mathbf{u}_0 = \begin{Bmatrix} -0.183 - 0.025j \\ -0.153 + 0.4348j \end{Bmatrix} \text{mm}$$

For the second test

$$\mathbf{f}_2 = \begin{Bmatrix} 0 \\ 0.1 \times 0.15 \end{Bmatrix} \text{kg m}$$

and the response is

$$\mathbf{u}_2 = \begin{Bmatrix} u_{21} \\ u_{22} \end{Bmatrix} = \begin{Bmatrix} 0.45 \times e^{j \times 50 \times \pi/180} \\ 0.4 \times e^{j \times 300 \times \pi/180} \end{Bmatrix} = \begin{Bmatrix} 0.289 + 0.345j \\ 0.2 - 0.346j \end{Bmatrix} \text{ mm}$$

The difference is given by

$$\Delta \mathbf{u}_2 = \mathbf{u}_2 - \mathbf{u}_0 = \begin{Bmatrix} 0.489 - 0.0017j \\ 0.350 - 0.087j \end{Bmatrix} \text{ mm}$$

Then from Equation 4.36, the imbalance is calculated as

$$\mathbf{f}_0 = \begin{bmatrix} 0.015 & 0 \\ 0 & 0.015 \end{bmatrix} \begin{bmatrix} -0.183 - 0.025j & 0.489 - 0.002j \\ -0.153 + 0.435j & 0.350 - 0.087j \end{bmatrix}^{-1} \begin{Bmatrix} -0.2 + 0.346j \\ -0.150 - 0.250j \end{Bmatrix} \text{ kg m}$$

This simplifies to give

$$f_0 = \begin{Bmatrix} -0.0192 + 0.0032j \\ -0.0135 + 0.0108j \end{Bmatrix}$$

Hence

$$|f_0| = \begin{Bmatrix} 0.0195 \\ 0.0173 \end{Bmatrix} \text{ kg m}$$

The phases are given by

$$\theta_1 = \tan^{-1}\left(-\frac{0.0032}{0.0192}\right) \qquad \theta_2 = \tan^{-1}\left(-\frac{0.0108}{0.0135}\right)$$

So the corresponding phases are 171° on plane 1 and 141° on plane 2.

There are several interesting observations to be made at this point. In applying this two-plane approach, inherently this means that two degrees of freedom have been adjusted. In the case of a rigid rotor, there are only two modes (in each direction). This implies that for a rigid rotor, if it is two-plane-balanced, it can be run at any speed. This is not true for a flexible rotor, and hence, it is important to understand this distinction and its relevance to specific plant items.

The issues involved are now illustrated by reference to the rigid rotor shown in Figure 4.8. The parameters are shown in Table 4.2, and a rundown plot is shown in Figure 4.9.

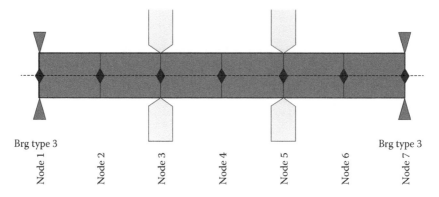

FIGURE 4.8
A rigid rotor example.

TABLE 4.2

Parameters of the Model

Distance between bearings (m)	1.5
Shaft diameter (m)	0.1
Bearing stiffness (N/m)	10^7
Bearing damping (Ns/m)	10,000
Disc thickness (m)	0.1
Disc diameter (m)	0.3
Material density (kg/m³)	7800
Young's modulus (N/m²)	2.1×10^{11}

The Campbell diagram for this system is shown in Figure 4.10 which illustrates that there is a minor gyroscopic influence on this machine, and the rundown plot shows peaks at 2200 and 6050 rev/min.

Some interesting features can be seen immediately from Figure 4.9. Most importantly, the response seen at the two bearings is almost identical at low speed and only begins to diverge above the first critical speed. This should come as no surprise as the dynamic behaviour in the lower speed range will be dominated by the first mode which is symmetric with respect to the two bearings. Suppose that this machine is to be run around 2000 rev/min. This is below the first critical speed and it may be anticipated that single-plane balancing would suffice. Figure 4.11 shows the behaviour of this rotor after single-plane balancing at 2340 rev/min. The balancing was achieved using a single trial mass and a single correction mass. It is clear that although the vibration at bearing 1 has been reduced to zero at the balancing speed of 2340 rev/min, at higher speed the amplitude has increased, but note that this takes no account of changes at bearing 2, or indeed, anywhere else on the machine.

Above the first critical speed, the vibration at a single speed and a single location can still be reduced by single-plane balancing, but the vibration

FIGURE 4.9
'As-found' response.

FIGURE 4.10
Campbell diagram.

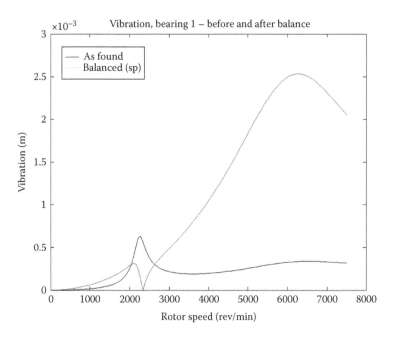

FIGURE 4.11
Before and after balancing with a single mass. *Note:* 'sp' denotes single plane balance.

elsewhere on the machine may increase. Figure 4.11 shows that at the point of measurement, the vibration can be reduced at a specific speed by single-plane balancing, but the effects at all other speeds are unpredictable without detailed knowledge of the machine's dynamic behaviour. In fact, a rather more satisfactory result may be obtained even with this simple single degree-of-freedom approach, simply by recognising the modal behaviour of the system.

In Figure 4.11, at speeds above around 3000 rev/min, the vibration becomes dominated by the second mode, which in this case we know to be a 'rocking' motion. In Figure 4.12, the calculated correction mass has been divided equally between the two discs, hence negating any chance in excitation of the second mode. The difference in behaviour is striking.

It is noticeable, however, that Figure 4.12 has not shown a reduction of the vibration level to zero at the balancing speed. This is because the calculations were based on a single trial mass whilst the correction involved two added masses: in other words, the modal content was different. The procedure may be refined by the use of a trial mass at the same position on each of the two discs, and the result of this procedure is illustrated in Figure 4.12.

This approach is very similar to modal balancing which is discussed in connection with flexible rotors. In the current context, dealing with rigid rotors, two-plane balancing is, in general, the most common method of

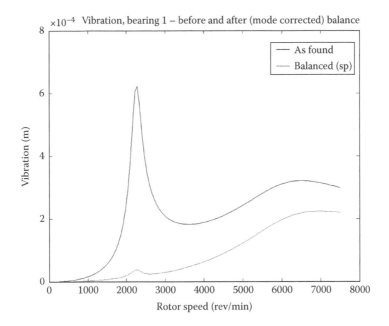

FIGURE 4.12
Balancing with two equal masses.

balancing although, as shown earlier, single-plane methods are appropriate in some circumstances. The examples shown earlier illustrate the importance of recognising a system's modal behaviour to obtain the best results from a balancing procedure, and two-plane methods implicitly recognise the two modes of a rigid body. Consequently, if a rigid rotor is two-plane-balanced at any speed, it will be balanced for all speeds. It is easily shown that two-plane balancing reduces the vibration to zero across the speed range. There are, of course, a number of caveats which should be mentioned here. The sensitivity of balancing will vary with rotor speed and little care is needed to make the most appropriate choice. Secondly, all balancing assumes linearity, a condition which is not strictly satisfied in practice, and some iteration may be required to obtain optimum balance conditions. Finally, no rotor is truly rigid, and this is particularly important with larger machines: the degree to which the assumption of rigidity is valid should be guided by the ratio of the operating speed to the lowest flexural free-free natural frequency.

Note that in Figure 4.13, the vibration is reduced to zero at a single speed at which the single-plane correction was calculated and applied. This perhaps needs a little clarification: trial masses were added equally to the two discs and the corresponding correction masses were calculated using the single-plane procedure. This relies on a single measurement point normally, but not necessarily, a bearing pedestal.

FIGURE 4.13
An example of balancing at a single speed.

The two-plane procedure utilises measurements from two distinct locations and is consequently able to take due recognition of the modal contributions to the vibration signal. Since a rigid rotor has only two modes of vibration (in each direction), the full complexity of the motion can be dissected, and this is why a rigid rotor balanced in two planes is balanced for all speeds.

So far the discussion has focused on rigid rotors, yet much of it is directly applicable to flexible rotors also. A rigid rotor can operate over any speed range after two-plane balancing because both (i.e. all) of its modes of vibration have been balanced. But a flexible rotor has more than two modes: many machines have only two modes within the running range, but, very often, the third mode is only a little higher in frequency and consequently has significant influence on the operation. The implication of this is that two-plane balancing will only yield satisfactory results at a specific rotational speed (or at best a narrow range of speeds). With this caveat, however, two-plane balancing can be an effective procedure for flexible rotors, and the approach presented earlier is called the influence coefficient method.

The influence coefficient approach, as before, involves two measurement points, two trial masses (involving two separate locations) and two trial runs at the same speed as the original condition. Whilst the implementation of this approach is exactly the same as in the rigid rotor case, it is important to appreciate the difference in behaviour of a flexible rotor. For instance, whilst ideal balance may be attainable at the running speed, the configuration yielding this condition may give an unacceptable vibration level as the machine goes through critical speeds during transient operation. Hence, it may be necessary to adopt some compromise between the requirements over the speed range as a whole.

4.3.3 Modal Balancing

A rather different method of balancing of a flexible rotor is the modal approach in which balancing is undertaken at several speeds, normally as close as possible to the critical speeds of the system. In principle, the balancing speeds chosen should be the critical speeds, but sometimes this is not possible owing to unacceptably high vibration levels. A single measurement point is required, but the number of balance planes used equals the number of modes to be balanced, normally two or three. Unlike the influence coefficient method, however, modal balancing requires prior knowledge of the machine in the form of the mode shapes and corresponding critical speeds. This knowledge may be derived from a theoretical model or previous operational data.

The procedure amounts to a multiple application of the single-plane approach, except that at each speed, trial masses must be added to each of the balance planes in proportion to the mode shape corresponding to the speed being studied. The premise of the approach is that each mode is independent of all others (a point discussed later), and hence the response at one critical speed is dominated by the corresponding mode and consequently each mode can be balanced independently. This is now illustrated with a simple example. It is known that critical speeds occur at 1200 and 2000 rev/min, respectively; the two balance planes are on two discs which are symmetric about the centre line as shown in Figure 4.8. The two mode shapes take the form $\begin{Bmatrix} 1 \\ 1 \end{Bmatrix}$ and $\begin{Bmatrix} 1 \\ -1 \end{Bmatrix}$. After a trial run at 1200 rev/min (or as close to it as the machine can safely run), the calculated correction is 40 g at 30° ahead of the phase reference and at a radius of 100 mm. Equal correction masses are then attached to the machine and the speed is then increased to 2000 rev/min, and after noting the initial vibration reading, equal trial masses are added in anti-phase to the two balance planes (at the same radius of 10 cm). From this the required correction is computed to be 20 g at 45° after the phase reference. Table 4.3 shows the magnitude and positions of the correction masses at 100 mm radius. The four masses will balance imbalance excitation of the first two modes and in many circumstances this will yield satisfactory operation. Only rarely do machines in current service require balancing for higher modes and in such cases more balance planes are required. Of course, on

TABLE 4.3

Modal Balancing Example

	Plane 1		Plane 2	
	Mag (g@100 mm)	Phase	Mag (g@100 mm)	Phase
Mode 1	40	30	40	30
Mode 2	20	−45	20	135
Resultant	49	7	40	59

each of the two planes in this example, the two masses may be replaced by a single one representing their (vector) sum, but in many cases, this may not be convenient as it would require an additional rundown of the machine.

Modal balancing is more satisfactory for machines which operate over a range of rotor speeds. The concepts were introduced around 1980 by Parkinson et al. (1980), yet the influence coefficient approach is still the more common method in industry. Both approaches rely on the concept of linearity and both methods can be easily applied in a routine approach, yet there are some pitfalls of which plant operators must be aware. One of the most common reasons for inconsistencies in balancing procedures is the presence of a bend. The behaviour of bent rotors rises to a synchronous vibration; it does not behave in the same way as imbalance because the variation of force with the rotor speed is quite different and the implications of this are quite profound. It is discussed in detail in Section 4.3, but the essential point to note here is that although a bend gives a synchronous response, its variation with speed is quite different to that of imbalance. In essence, forcing due to a bend is constant as opposed to the dependence on speed squared for imbalance.

In fact, a bend in a rotor can be easily diagnosed by the presence of significant vibration at very low speed (i.e. far below the first natural frequency). As explained in Section 4.3, a bend can only be compensated by a balance mass for a single speed, and consequently, provided a machine is to operate only at one speed, it may be balanced without regard to the bend. In such cases, however, due care must be taken to ensure that vibration levels are acceptable during run-up and rundown.

An alternative procedure is available for machines which are required to operate over a range of rotor speeds. In this case, the bend must be first 'measured' by observing the response at low speed. These levels are subtracted (vectorially) from the measure response in balancing with correction masses calculated on this basis.

To examine the effect of a possible bend, let us return to the example quoted on pages 98–100, but with the added proviso that at very low speed, the response was measured as 10 μm, 30° ahead of the phase reference. The question now is how to best balance the rotor. In the earlier example, the required correction balance was calculated as 9.03×10^{-4} kg at 134°. If the rotor is to be run only at 1500 rev/min, then the presence of the bend may be ignored, but at any other speed the rotor will not be correctly balanced. However, there is a danger in this approach as the machine may go through very high vibration during rundown.

If, on the other hand, the machine is required to operate over a range of speed, then a compromise balance procedure should be employed. The basis of this is to use the difference between the vibration reading and the bend in place of the vibration reading itself. Hence,

$$me = -0.001 \times \left(\frac{25}{\sqrt{2}}(1+j) - 10\left(\frac{\sqrt{3}}{2} + \frac{j}{2} \right) \right) \times \frac{1}{10\sqrt{3} - \frac{25}{\sqrt{2}} + j\left(-10 - \frac{25}{\sqrt{2}} \right)}$$

This imbalance may be evaluated as $me = (4.6217 - 3.1984j) \times 10^{-4}$ kg m. Note that this is a quite different imbalance to that calculated without the bend, and the application of these diverse answers demands a little insight. This corresponds to an imbalance magnitude of 5.62×10^{-4} kg m at $-35°$. If this rotor is to be run only at the balancing speed (1500 rev/min), then the lowest vibration will be attained using the balance procedure ignoring the bend. On the other hand, the rotor is to be run at a range of speeds, it can be shown that making allowance for the bend, as in this example, will limit the overall vibration to the magnitude of the bend. The reason for this slightly complicated behaviour is that the response of bends and unbalance is rather different even though they both give rise to synchronous (once per revolution) vibration. This is explained in Section 4.4.

The underlying concept of this approach is clear; by subtracting the bend, the operator is isolating that part of the vibration which is caused by imbalance of the proceeding accordingly.

The nature of oil journal bearings also leads to some problems with regard to imbalance response and the balancing process. The problem of non-linearity is discussed in Section 4.3.4 and can be effectively overcome by balancing in an iterative manner, but often this is not necessary. A rather more profound problem is that the bearings provide an overwhelming proportion of the system damping which means that the damping is not 'classical', which in turn means that the mode shapes are complex and that they are not orthogonal. The important point here is that the lack of orthogonality implies that some of the fundamental requirements of modal balancing are not strictly satisfied. An enhanced procedure, called bimodal balancing, has been suggested by Kellenburger and Rihak (1988), but to the best of the author's knowledge, this has yet to be applied in an industrial setting. In fact, despite the theoretical doubts, the evidence suggests that acceptable modal balancing can be obtained on machine mounted on oil journal bearings.

It is noted here that the importance of the distinction between rigid and flexible rotors becomes clear. A rigid rotor, balanced in two planes, will be balanced for all speeds. This is clear in modal terms since all the modes (i.e. two rigid body modes) have been balanced. With a flexible rotor however, this will not be the case as other modes assume importance with increasing rotor speed. Of course, there is no such thing as a truly rigid rotor, but the criterion is based on the closeness of the first flexural free-free natural frequency to the running range. The state of balance of a flexible rotor must be considered at its speed of operation.

4.3.4 Non-Linear Effects

On many machines, in particular those mounted on oil film journal bearings, there are some non-linear effects. To explore this, the system described in Table 4.1 is considered with the imbalance increase by the factors specified in Table 4.1. Table 4.4 shows the amplitude (relative to the bearing clearance) at several speeds for a range of values of imbalance. The scale given in the table is a multiplier applied to the imbalance pattern as in Table 4.1. The calculations have been carried out using the short bearing theory: whilst this has a number of assumptions and is not precise, it is representative of many of the phenomena observed in oil film bearings. The feature of note here is the inherent non-linearity. At 900 rev/min, the non-linear behaviour only becomes apparent as the imbalance increases. More precisely we note that the non-linear behaviour becomes important when the displacement at the bearing is commensurate with the clearance which would seem to be physically reasonable. Note from the table that at higher speeds this characteristic appears at lower values of imbalance.

At this point consider the effect this has on the balancing procedure: it is immediately clear that the response, in all cases, shown in Table 4.4 increases less fast than the applied imbalance. This behaviour may be characterised by bearing which effectively 'stiffens' with increasing amplitude. Of course, this has direct relevance to balancing since by the application of trail masses, the sensitivity of the system is established prior to the application of the appropriate correction masses. A consequence of this apparent stiffening of the bearings is that the first estimate of the required correction will be, in general, an underestimate, but it will yield reduced vibration levels enabling

TABLE 4.4

Amplitude Range Relative to Clearance

Imbalance Scale	900 rev/min	1800 rev/min	2700 rev/min
1	0.054	0.155	0.51
2	0.106	0.299	0.826
4	0.209	0.539	1.15
8	0.394	0.84	1.40
16	0.672	1.12	1.58

a second balancing to be undertaken. This may not be necessary in all cases but is dependent on the extent of the 'as-found' amplitudes and the required tolerances. Where two-stage balancing is undertaken, it yields valuable practical data on the non-linear bearing behaviour.

4.3.5 Recent Developments

Over recent years, a number of authors have proposed methods for balancing in a single step (i.e. without any 'trial runs'). In an industrial setting such a technique could prove very valuable. For a particular machine, given extensive and detailed records, an engineer may well have good insight into the particular machine's sensitivity and such prior information will help reduce the need for test runs.

Novel approaches are based on system identification techniques which are discussed at some length in Chapter 7. The estimation of imbalance in this way is a particular case of the fitting of a number of parameters of a numerical model to measured data. In this case, a vector of imbalances at a predetermined set of balance plane is included as parameters to be identified (Lees and Friswell 1997). The technique is still relatively new, but laboratory trial and limited site application have given very promising results. Chapter 7 gives details.

4.4 Rotor Bends

Section 4.3 gives an account of mass unbalance, and this is undoubtedly the most prevalent 'fault' condition in rotating machinery. It is rather doubtful whether imbalance is properly termed as a machine fault as all machines have it to some degree. What is a machine fault is an excessive level of unbalance when judged against some suitable criteria. As discussed in Section 4.3, imbalance always gives rise to synchronous vibration but the converse is not true, that is a once-per-revolution signal is not necessarily the result of imbalance. It may be that the shaft has a bend. A bend, like unbalance, gives rise to synchronous vibration, but there is some significant difference in the characteristics of an unbalanced rotor and one which is bent: of course, a real rotor may be unbalanced and bent at the same time, but it is convenient to separate the two effects to illustrate the principles.

First consider the rotor shown in Figure 4.14.

FIGURE 4.14
A bent rotor.

The figure depicts a simply rotor with a bend near the centre. It is important, however, to clarify what is meant by the term bend. What is being discussed here is quite different from the catenary of a flexible rotor which is discussed in Chapter 5 in the study of misalignment. In the case of a bend, the distorted shape of the rotor rotates about the centre line of the shaft, whereas the catenary is fixed in space and time, at least for the case of an axisymmetric rotor.

The equation of motion of a rotor with a bend can be derived by means of Lagrange's equation. If the bend vector is **b**, then the potential energy of the rotor, with position vector {x}, is given by

$$U = \frac{1}{2}\{x - b\}^T[K][x - b]$$
(4.36)

The physical validity of this expression may be verified by observing that, in some coordinate system, a straight shaft could be arranged so that x = **0**, but if the rotor has a bend, that would not be a configuration with zero potential energy. This zero energy state can only be achieved when {x − b} = **0**. The bend **b** is just the shape the rotor would take up when resting on a frictionless table. It should be added that in many cases such an imaginary table would be very large!

The kinetic energy is straightforward and is given by

$$T = \frac{1}{2}\{\dot{x}\}^T[M]\{\dot{x}\}$$
(4.37)

A straightforward application of Lagrange's equation yields

$$[M]\{\ddot{x}\} + [K]\{x - b\} = 0$$
(4.38)

as the equation of motion for a bent shaft with no damping and no unbalance. It is instructive to examine the behaviour of this simplified case before proceeding to more complicated situations. Considering a rotor now with a steady angular speed of Ω, the vibration level (with the inclusion of a damping matrix **C**) may be obtained directly giving

$$x = [-M\Omega^2 + j\Omega C + K]^{-1} K b e^{j\Omega t}$$
(4.39)

Hence, it is clear that a bend in a rotor gives rise to a once-per-revolution excitation, but the amplitude variation of this with rotor speed is fundamentally different from that caused by rotor unbalance. This is clear from the absence of a term proportional to Ω^2 in the numerator of Equation 4.39. To illustrate this point, consider a rotor with an unbalanced distribution $\mathbf{M}\varepsilon$, where ε is the mass eccentricity and vibration level is

$$\mathbf{x}_{bal} = [\mathbf{K} + j\Omega\mathbf{C} - \mathbf{M}\Omega^2]^{-1}\mathbf{M}\varepsilon\Omega^2 e^{j\Omega t} \tag{4.40}$$

It is obvious that we can express this in terms of Fourier components such that

$$X_{bal}(\Omega) = [\mathbf{K} + j\Omega\mathbf{C} - \mathbf{M}\Omega^2]^{-1}\mathbf{M}\varepsilon\Omega^2 \tag{4.41}$$

For the sake of simplicity, we assume that damping is small and consequently the peak response occurs at $\Omega = \Omega_c$ given by the condition $|\mathbf{K} - \Omega_c^2\mathbf{M}| = 0$. At critical speed, the response due to bend and unbalance will be equal if

$$\mathbf{Kb} = \mathbf{M}\varepsilon\Omega_c^2 \tag{4.42}$$

Assuming this to be the case, let us now compare the displacements due to unbalance and bends at zero speed and high (infinite) speed. Directly from Equations 4.41 and 4.42, it is observed that

$$\begin{aligned} \mathbf{X}_{bal}(0) &= 0 & \underset{\Omega\to\infty}{\mathrm{Lim}}(\mathbf{X}_{bal}(\Omega)) &= -\varepsilon \\ \mathbf{X}_{bend}(0) &= \mathbf{b} & \underset{\Omega\to\infty}{\mathrm{Lim}}(\mathbf{X}_{bal}(\Omega)) &= 0 \end{aligned} \tag{4.43}$$

Figure 4.15 shows a rotor model with unbalance distributed on the two discs. The response of this system with a unit unbalance on the disc is shown together with a simple bend having the same response at resonance. The comparison of the two curves illustrates the very important point that whilst bends and unbalance both give rise to synchronous signals (once per revolution), the variation as rotor speed varies is quite different. A consequence of this is that, although the effects of a bend can be compensated by balancing, this can only be done at a single rotor speed.

FIGURE 4.15
Basic layout of test model.

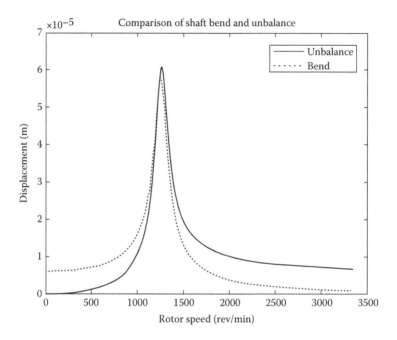

FIGURE 4.16
Imbalance and bend responses.

In Figure 4.16, a bend has been introduced, the force of which exactly matches that from the unbalance distribution. Notice that the unbalance and bend components match at this single frequency (21 Hz), but nowhere else.

Note that below 21 Hz, the unbalance response is below that of the bend whilst the opposite is the case at higher frequencies.

Whilst this argument has been explained implicitly using a single degree-of-freedom system, it is equally valid for a more complex system. In this subsection the only assumption (apart from linearity) is classical damping so that modes are real. It is now assumed that both unbalance and a bend are present (ignoring damping for clarity):

$$\mathbf{M}\{\ddot{x}+\ddot{\varepsilon}\}+\mathbf{K}\{x-b\}=0 \tag{4.44}$$

so that at constant speed Ω, the equation of motion becomes

$$[\mathbf{K}-\mathbf{M}\Omega^2]x = \mathbf{M}\Omega^2\varepsilon+\mathbf{K}b \tag{4.45}$$

This equation can be solved either directly or in modal terms.

Consider now the rotor shown in Figure 4.15. Two discs are mounted on a uniform shaft and there is a bend offset from the centre. In this example the bearings are considered to be rigid, although this is not pertinent to the case. With no unbalance on the rotor, the vibration at the centre is shown

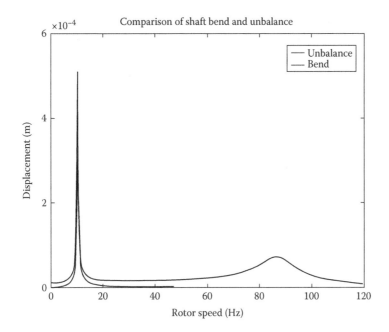

FIGURE 4.17
Response to a bend and imbalance.

in Figure 4.16. The frequency range has been deliberately chosen to include only the first mode (excitation at higher speeds is considered later). Imposed on this graph is the vibration arising from a pure imbalance that has been chosen to give the same vibration at resonance.

Before leaving this topic, consider a system in which the second mode is within the running range. The response due to both a bend and unbalance distribution is shown in Figure 4.17.

In this case, the model has been arranged to have equal response to unbalance and the bend at the first resonance (in this case around 10 Hz); at the second peak, the response of the bend is negligible. Of course, one could have a bend to give a response at the second peak – but one cannot have both.

As is clear from Equation 4.45, a bent rotor can only be 'balanced out' at a single speed.

4.5 Concluding Remarks

This chapter has discussed the topics of imbalance and rotor bends. The two are closely associated as they both give rise to synchronous vibration signals, but they are rather different in the variation with rotor speed. An important consequence of this is that a rotor bend can only be compensated at a single speed. This can be very useful in practice, but it is important to appreciate the limitation to a single speed.

A discussion has been given of the description of imbalance and other excitation in terms of the natural frequencies and mode shapes of the system. After some manipulation this yields an elegant formulation of the machine's response. This is important, not for the sake of academic niceties, but rather that this formulation clarifies the nature of the response. Further discussion of the use of mode shapes is given in Section 3.5.

Single- and two-plane balancing have been outlined and applied to both rigid and flexible rotors. Although similar approaches are employed, it is important to appreciate the distinction between rigid and flexible rotors. A rigid rotor, balanced in two planes, will be well balanced at all speeds. This is because all modes have (implicitly) been balanced. This is not the case with a flexible rotor which must be balanced at its operational speed.

Problems

4.1 A fan operates at 1500 rev/min, just above its first critical speed, and it is to be balanced using a single-plane approach. Using data from one bearing in a single direction, it is observed that at running speed the vibration level is 5 μm at 60° ahead of the shaft reference. A 10 g mass is attached to the shaft at a radius of 200 mm, 180° from the marker, and this gives vibration at a running speed of 3 μm at 30° ahead of the reference. Calculate the position and magnitude of the required balance mass.

4.2 On the same fan at a very low speed, the vibration level is 2 μm, 60° behind the reference. Discuss the arguments for changing the balance and calculate any revision to the required mass and position and calculate the expected vibration level at 1500 rev/min.

4.3 An alternator rotor operates at 3000 rev/min, above its second critical speed, and must be balanced in two planes. The vibration readings 'as found' and with two trial masses are shown in the table.

	Bearing 1		Bearing 2	
	Amplitude (mm)	Phase (°)	Amplitude (mm)	Phase (°)
As found	0.35	80	0.22	−40
100 g at 0°, 200 mm radius Disc 1	0.20	45	0.30	30
100 g at 0°, 200 mm radius Disc 2	0.30	30	0.10	0

Calculate the magnitude and position of the required balance masses.

4.4 The rotor in Question 3 is found to have a bend. When rotating at 50 rev/min, it is observed to vibrate with amplitudes of 50 and 100 μm, respectively, at the two bearings. Should the balance be modified? If so, what will be the vibration at 3000 rev/min with the new balance configuration?

4.5 A machine operating over a broad speed range is to be balanced modally. Explain why this is the most appropriate approach. It is known that, referring to the two balance planes, the two modes are $\{1\ 1\}^T$ and $\{1\ -1\}^T$. Measurements are made on the cap of bearing 1. The vibration 'as found' on bearing 1 is 50 μm at 30° ahead of the phase reference. Close to the first critical speed, with 50 g masses attached to both discs at the phase reference mark and 100 mm radius, the vibration is 35 μm at −45°. The mass on the second disc is then moved by 180° (at the same radius) and the machine is then run up to its second critical speed. The vibration is now measured as 20 μm at 60°. Determine the magnitude and position of the appropriate balance masses.

References

Foiles, W.C., Allaire, P.E. and Gunter, E.J., 1998, Review: Rotor balancing, *Shock and Vibration*, 5, 325–336.

Friswell, M.I., Penny, J.E.T., Garvey, S.D. and Lees, A.W., 2010, *Dynamics of Rotating Machines*, Cambridge University Press, New York.

Kellenburger, W. and Rihak, P., 1988, Bimodal (complex) balancing of large turbo-generator rotors having large or small unbalance, in *Institution of Mechanical Engineers Conference on Vibrations in Rotating Machinery*, Edinburgh, U.K., Paper 292/88.

Lees, A.W. and Friswell, M.I., 1997, The evaluation of rotor imbalance in flexibly mounted machines, *Journal of Sound and Vibration*, 208(5), 671–683.

Parkinson, A.G., 1991, Balancing of rotating machines, *Proceedings of the Institution of Mechanical Engineers, Part C, Journal of Mechanical Engineering Science*, 205, 53–66.

Parkinson, A.G., Darlow, M.A. and Smalley, A.J., 1980, A theoretical introduction to the development of a unified approach to flexible rotor balancing, *Journal of Sound and Vibration*, 68, 489–506.

Reiger, N.F., 1986, *Balancing of Rigid and Flexible Rotors*, The Shock and Vibration Information Center, United States Department of Defense, Washington.

5

Faults in Machines – Part 2

5.1 Introduction

Chapter 4 discusses the important issues arising with mass imbalance and the ways in which this can be alleviated by balancing. Rotor bends are also analysed: this is a fault which is often confused with imbalance, and in some instances, it may be appropriate to correct a bend by balancing, but in general, it is important to recognise the essential differences. In this chapter, some further 'common' faults in rotating machine are discussed. Whilst these are generally less common than imbalance (the most prevalent single fault), they tend to be somewhat more complex. Furthermore, faults interact and it sometimes becomes difficult to identify the root cause with any degree of certainty.

The chapter begins with a discussion of misalignment, which is in itself a complicated issue that remains an active area for researchers as there is not yet a satisfactory overarching theoretical foundation which supports the variety of observed behaviour. After describing the basic features, some recent thoughts on the topic are outlined. Another important fault is cracked rotors. Whilst (thankfully) much less common than other defects outlined, the consequences can be disastrous both in terms of plant and a threat to the life of operators. Torsion and displacement controlled problems are discussed.

The last section of the chapter deals with a discussion of the various ways in which the faults interact. It is these interactions that sometimes make it difficult to ascribe a prime cause to a machine defect. So many of these effects become inter-related and this is, perhaps, the main reason why automated detection systems have made only limited progress despite numerous attempts.

5.2 Misalignment

Rotor misalignment is second in importance only to mass imbalance in the list of significant faults on rotating machinery. Despite its major industrial significance, it is, as yet, incompletely understood, and later in this section, some current ideas are discussed. The incomplete understanding to date is reflected in the lack of research publications in this area: more than a 1500 papers are readily available for mass imbalance, whereas for misalignment, one has difficulty in finding more than 60. It is interesting to reflect on why this should be so and it becomes clear that this field of study is indeed complex and fraught with difficulty.

The simplest case to consider is a rigid shaft mounted on two bearings. Consideration of moments on the system will dictate that appropriate loadings fall on the two bearings, but if they are not correctly aligned, the loads will not be taken evenly across the face of the bearing and rapid wear may be the result. Depending on the type of bearing used, substantial deviations in bearing performance can be observed, but the true situation is significantly more complex insofar as the operation of the complete machine may be significantly impaired. Depending on the type of machine in question, there will be fine clearances within it through which working fluid flows, and hence any change in the shaft position within these clearances, will have a profound effect on machine operation. So even in the simplest case, it is important to ensure that (1) the shaft and the bearings are co-axial and (2) the shaft passes axially through internal clearances and neck rings. This latter influence is touched upon in Chapter 6. Where machines of this type have to be coupled together (e.g. an electric motor and a pump), it is commonly achieved with a flexible coupling.

With large machines, however, the coupling between rotors is usually (nominally) rigid. For instance a large steam turbo-generator may comprise seven rotors each mounted on two bearings. These seven would be high-pressure (HP) turbine, intermediate-pressure (IP) turbine, three low-pressure (LP) turbines, the alternator and the exciter. These large units are typically 50 m in length with a total rotor mass of about 250 t. This is just an example and the combination of rotors will vary with the design of unit. These rotors will operate at widely different temperatures, and the need to maintain the complete unit in the correct configuration presents an engineering challenge.

Figure 5.1 illustrates the two types of misalignment in which the abutting rotor faces can meet with lateral or angular relative displacement. In reality, most cases will be a combination of these two as shown in the third diagram. It is, however, convenient to view parallel and angular misalignments separately as some distinct effects arise.

5.2.1 Key Phenomena

It is very common for a system to display the geometries illustrated in Figure 5.1. The mismatch can be overcome in three ways:

  a. The use of a flexible coupling which allows for this mismatch but transmits the torque. These devices come in a range of types, but most involve some non-linear behaviour. Flexible couplings are used

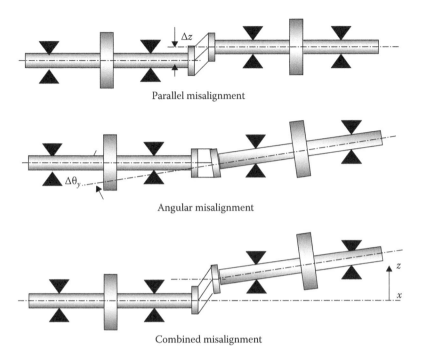

Parallel misalignment

Angular misalignment

Combined misalignment

FIGURE 5.1
Types of misalignment. (Reproduced from Sinha, J.K. et al., *J. Sound Vibrat.*, 272(3–5), 967, 2004. With permission.)

predominantly on smaller machines, but a full discussion of their (interesting) properties is outside the scope of this chapter.

b. The location of adjacent bearings is changed in such a way as to substantially reduce the mismatch, both laterally and angularly.

c. Under some circumstances, it may be necessary to apply forces and moments to the coupling to force a rigid connection. This should be a last resort as it is always necessary to minimise the forces and moments applied. Some of the possible consequences are discussed in the following.

The ways of correcting misalignment in flexible rotor are addressed in Section 5.2.3, but the current discussion focuses on the effects observed due to misalignment. The primary cause of difficulty is that the anomalous forces and moments at the couplings cause a redistribution of (usually inappropriate) bearing loads and this has numerous consequences:

1. Because the bearing (static) loads change the bearing stiffness and damping, the natural frequencies will change.

2. This may give rise to stability problems or bearing overload.

3. Clearances within the machine will change as the shaft takes up a new position: this has consequences on machine performance. Together, these influences may lead to a rub, with numerous consequential issues.

4. Because the shafts now rotate about an axis which is not its true geometric or mass centre, the effective imbalance may alter.

It should be emphasised that the crucial aspect of alignment is the apportioning of correct load on all bearings and minimisation of rotor stresses. Excessive bearing loads can lead to rapid wear, but in many instances, unduly low loading can lead to even greater difficulties. This problem arises in many types of oil journal bearings, often used in large machines owing to their good load capacity. However, they tend to become unstable if the (static) load is inadequate. Under these conditions, the oil flow within the bearing tends to break up and rotate en-block around the rotor. The phenomenon is known as oil whirl, and this gives rise to appreciable vibrations at speeds up to half the shaft rotation speed (see, e.g. Muszynska 2005). Using energy considerations, it can be shown that this phenomenon can never exceed half the shaft speed and for that (slightly illogical) reason, it is often known as 'half-speed whirl'. It can be a very dangerous condition, but there is an even more serious variant known as oil whip. Whilst this is fundamentally the same phenomenon, the distinguishing feature is that the vibration 'locks in' to a free-free mode of the rotor system which is below half shaft speed. The bearing stiffness and damping become effectively zero, and consequently, very high vibration levels arise which can be extremely dangerous in terms of machine integrity and safety of operational staff.

There are essentially two ways to resolve this difficulty: the first is to connect the mating shafts with a flexible coupling which is a device that transmits torque whilst having a low flexural stiffness. Flexible couplings have some associated problems but can provide a very effective means of connecting machine components. The basics of this approach are discussed in Section 5.2.2.

The alternative approach is a rigid coupling of the two rotors, and this is employed in most large machines. Whilst avoiding some of the problems associated with flexible couplings, it does lead to other difficulties as outlined in Section 5.2.3.

5.2.2 Flexible Couplings

Machines may be coupled using a variety of mechanisms, but we will illustrate some of the modelling problems with reference to a single membrane coupling, a type which is in common use. The principle is illustrated in Figure 5.2. Each of the two shafts is connected to different parts of a flexible membrane which becomes the only component through which the rotors are connected. Hence, torque is transmitted, but bending moments are not, because the membrane can bend. This amounts effectively to the insertion

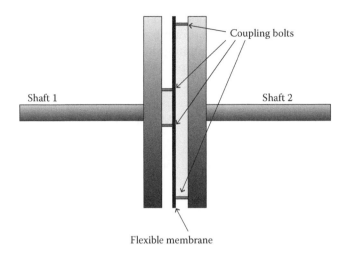

FIGURE 5.2
A membrane flexible coupling.

of a hinge in the rotor. In practice, pairs of couplings are often used, separated by a spacer shaft to obtain better vibration isolation. Here however,
the modelling difficulties are discussed with reference to a single coupling.

Let us now consider the model of a turbine and alternator discussed earlier, but with a single flexible coupling between them. Note that although
flexible couplings have been used predominantly on smaller machines, their
application is now widening to encompass large plant items. We begin by
modelling the two rotors separately; the lowest natural frequencies of both
rotors are shown in Table 5.1. The nodal co-ordinates at the end of rotor 1
are $\{u_{1n} \quad v_{1n} \quad \theta_{1n} \quad \psi_{1n}\}^T$ whilst those for the connected end of rotor 2 are
given by $\{u_{21} \quad v_{21} \quad \theta_{21} \quad \psi_{21}\}^T$. The coupling is modelled by applying the
constraints $u_{1n} = u_{21}$, $v_{1n} = v_{21}$. In this way, the two adjoining nodes, which
before connection had eight associated degrees of freedom, now have six.
This is achieved by means of a transformation matrix as follows:

$$
\begin{Bmatrix} u_{1n} \\ v_{1n} \\ \theta_{1n} \\ \psi_{1n} \\ u_{21} \\ v_{21} \\ \theta_{21} \\ \psi_{21} \end{Bmatrix}
=
\begin{bmatrix}
1 & 0 & 0 & 0 & 0 & 0 \\
0 & 1 & 0 & 0 & 0 & 0 \\
0 & 0 & 1 & 0 & 0 & 0 \\
0 & 0 & 0 & 1 & 0 & 0 \\
1 & 0 & 0 & 0 & 0 & 0 \\
0 & 1 & 0 & 0 & 0 & 0 \\
0 & 0 & 0 & 0 & 1 & 0 \\
0 & 0 & 0 & 0 & 0 & 1
\end{bmatrix}
\begin{Bmatrix} u \\ v \\ \theta_{1n} \\ \psi_{1n} \\ \theta_{21} \\ \psi_{21} \end{Bmatrix}
\tag{5.1}
$$

This forms part of a global transformation matrix $\mathbf{T}$ which has the dimensions $N \times (N-2c)$, where N is the number of (unconstrained) degrees of freedom and c is the number of couplings in the system. Using the subscripts c and u to denote coupled and uncoupled respectively, the mass and stiffness matrices are given by

$$\mathbf{K}_c = \mathbf{T}^T \mathbf{K}_u \mathbf{T} \quad \mathbf{M}_c = \mathbf{T}^T \mathbf{M}_u \mathbf{T} \tag{5.2}$$

We now consider the influence of the coupling in an example.

Example 5.1

A simple machine is shown in Figure 5.3. The two rotors have overall lengths of 3 and 4 m, respectively, and diameters of 250 and 300 mm. Both the central rotor overhangs are 0.5 m long. All four bearings have constant stiffness of 5×10^6 N/m. The mass of the coupling itself has been neglected. Since the rotors are (almost) rigid, each span has been modelled with two elements and each overhang with a single element. The natural frequencies of the system have been calculated under three conditions, namely, uncoupled, coupling of only displacement and finally full coupling (of displacement and slope). The results for the first two modes in each case are shown in Table 5.1.

Note that the effects of the coupling is different on the two modes depending on the particular mode shape.

TABLE 5.1

Natural Frequencies (Hz) under Different Coupling (Example 5.1)

	Uncoupled		Hinge Coupling	Solid Coupling
	Rotor 1	Rotor 2	Hinge Coupling	Solid Coupling
Mode 1 (Hz)	13.47	10.06	10.10	13.47
Mode 2 (Hz)	21.62	16.31	13.47	19.65

FIGURE 5.3
Sample system.

Couplings play a crucial role in the dynamic behaviour of rotating machinery, and the brief description given here is just an overview of methods used to adequately model the situations. In reality, couplings have mass and inertia properties which must also be modelled: furthermore because the masses are sometimes considerable and they load the shaft some distance from the bearings, appreciable effects on dynamic behaviour can result.

There is a variety of flexible coupling types and it is not the aim here to give a comprehensive survey, but rather a representative sample of their operation together with some indication of some of the practical difficulties which arise. The example of Hooke's joint is not only important in itself but also provides an excellent introduction to important concepts of flexible couplings. In essence, the requirement of a coupling is to connect some, but not all, degrees of freedom: usually the requirement is to transmit torque but not lateral forces or bending moments. In its simplest form, Hooke's joint, or Carden shaft, connects the two rotors simply by two pivots, on each of the two rotors, set at right angles. Basic kinematics shows that if the angular velocity of the first rotor is held constant at Ω_1, and β is the angular mismatch between the rotors, then the velocity of the second rotor becomes

$$\Omega_2 = \frac{\Omega_1 \cos\beta}{1 - \sin^2\beta \cos^2 \Omega_1 t} \tag{5.3}$$

And the resulting angular acceleration is given by the derivative of this expression, namely

$$\dot{\Omega}_2 = \frac{\Omega_1^2 \cos\beta \sin^2\beta \sin 2\Omega_1 t}{\left(1 - \sin^2\beta \cos^2 \Omega_1 t\right)^2} \tag{5.4}$$

Taking as an example an input speed of 50 Hz (3000 rev/min) and an angular mismatch of 5°, the variation in output speed is shown in Figure 5.4.

However, it is clear that the speed variations are modest, amounting in this case to about ±1/2%. The angular accelerations are appreciable, and this will impose high stresses on components such as coupling bolts and membranes. This figure of course is for a coupling in perfect condition. A defective bolt or failed membrane will change the situation completely.

There are various types of coupling in operation but they fall into three main groups:

1. Membrane couplings
2. Gear couplings
3. Rubber-block couplings

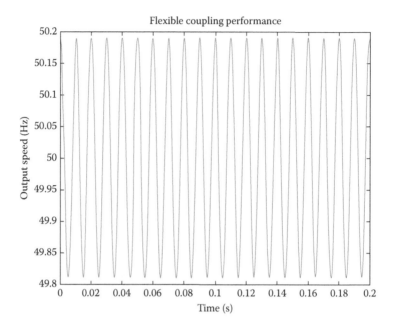

FIGURE 5.4
Speed variations.

There is also a fourth type, based on a fluid clutch principle, but these tend to be more specialised applications. Each type of coupling has its own virtues and faults. The gear type, for instance, has the benefit of being lighter than other types, but it is essential to maintain a good lubrication system. Balancing may also present some difficulties.

5.2.3 Solid Couplings

In large, high-power machines, such as turbo-generators, individual rotors are often rigidly connected. This is usually achieved by means of bolted flanges. To some extent, this is a simple situation to model, since the assembly of rotor can, in this case, be modelled as a single rotor with, of course, appropriate provision for the mass and inertia of the flange coupling. As discussed in Section 5.2.4, accurate modelling of these coupling is a little more complicated than as suggested by this simple formulation.

There are also some problems to examine the alignment of the overall machine. In selecting the position for the bearings, the choice is made based on two criteria:

1. Elimination of cyclic stresses in the shaft
2. Good distribution of bearing loads

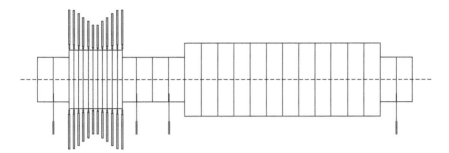

FIGURE 5.5
Idealised turbo-generator.

We illustrate the problem with a single-stage turbine that is connected to a large alternator (in practice, of course, there will be several turbine stages, but this has be simplified to a four-bearing arrangement for ease of illustration), as shown in Figure 5.5.

5.2.3.1 Catenary

The two rotors in this example are heavy and flexible, implying that both of them will `sag' due to gravity. The extent of this static defection is readily calculated using the mass and stiffness matrices for the system. Hence, we may write

$$\delta = \mathbf{K}^{-1}\mathbf{M}\mathbf{S}g \tag{5.5}$$

where $\mathbf{S} = \{0 \quad 1 \quad 0 \quad 0 \quad 0 \quad 1 \quad 0 \quad 0 \quad \cdots \quad \cdots \quad \cdots \quad \cdots \quad 0 \quad 1 \quad 0 \quad 0\}^{T}$ and g is the acceleration due to gravity. In the case considered, the turbine rotor is much stiffer than the long, relatively slender alternator. Assume that the bearings have been arranged to be in a straight line. This would appear to be a reasonable starting point in the absence of more information, but it soon becomes clear that such a simple arrangement is not satisfactory. To appreciate this, consider the catenary of the two rotors if they are not coupled. As shown in Figure 5.6, the two shafts have a mismatch both in terms of vertical position and the angle at which the mating faces meet. The two shafts fail to meet, and coupling the two in this configuration would induce high shear forces and bending moments in the shafts. Furthermore, the forces and moments change the distribution of bearing loads and hence alter the dynamic characteristics.

If the rotors are coupled in this configuration, the resulting shaft catenary is shown in Figure 5.7 where it will be noticed that although the maximum deflection has reduced slightly, there is very high curvature of the shaft near the coupling, indicating high bending moments (and, consequently,

FIGURE 5.6
Uncoupled rotors.

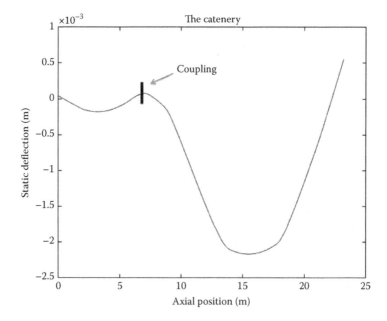

FIGURE 5.7
Rotors forced into coupling.

high stresses). Note also that these forces and moments would be stationary in space and therefore cyclic on the rotor, a fact that could (and occasionally has) lead to major problems including shaft failure.

Note however that in these figures, the vertical scale has been greatly magnified for the purpose of illustration. Setting aside for a moment the problems introduced by thermal differentials during operation, the problem of the basic layout can be resolved in a straightforward manner; in fact, there are several solutions to the problem.

For the present case, a suitable alignment may be reached in a two-stage process:

a. First the vertical level of bearings 2 and 3 is changed in such a way as to make the adjoining face parallel. This is actually quite straightforward to do as lowering/raising of the bearing simply changes the slope of each rotor and therefore, the end face. The heights are altered using shims under the bearing housings.

b. Having obtained parallel faces, the heights of either bearings 1 and 2, or 3 and 4 are changed in such a way as to ensure that at the junction, the faces are both parallel and concentric.

Both the steps are a matter of geometry and given the dimensions of the machine, the calculation is easily completed.

In the case of a solidly coupled system, common in large machines, the basic parameters required are the slopes of the neighbouring rotors in the uncouple state. In the current, admittedly rather simplified example, the axial positions of the bearings (as measured from the end of the rotor) are

Bearing 1 $z_1 = 1\,\mathrm{m}$
Bearing 2 $z_2 = 6.2\,\mathrm{m}$
Bearing 3 $z_3 = 7.2\,\mathrm{m}$
Bearing 4 $z_4 = 22.2\,\mathrm{m}$

And the coupling is midway between bearings 2 and 3 at $z_c = 6.7\,\mathrm{m}$.

The bearings are all initially in line and the slopes and heights of the two faces are as calculated in Figure 5.6. On real plant, they can be readily measured with the angular discrepancies that are quantified using feeler gauges. In this case, the displacement and slope at the couplings were calculated as

Rotor 1 displacement 0.1235 mm slope $sl_1 = -0.1267 \times 10^{-3}$
Rotor 2 displacement -0.0047 mm slope $sl_2 = +0.6649 \times 10^{-3}$

The aim now is to position the rotors so that their coupling faces (cross sections) are parallel. This may be achieved by lowering bearing 2 by $sl_1 \times (z_2 - z_1)$, giving a movement of 0.659 mm (δ_2). Similarly, bearing 3 should be lowered by $sl_2 \times (z_4 - z_3) = 9.31$ mm (δ_3). Since the appropriate angles have been used, the two faces will now be parallel, but further shifts are needed to obtain the correct displacements.

The vertical position of the first shaft at the coupling will now be

$$y_{c1} = 0.1235 - sl_1 \times (z_c - z_1)$$

Similarly, the position of the second rotor will be given by

$$y_{c2} = -0.0047 - sl_2 \times (z_4 - z_c)$$

Having calculated these values, a workable configuration is given by

Brg1   move by $y_{c1} - y_{c2}$
Brg2   move by $y_{c1} - y_{c2} + \delta_2$
Brg3   move δ_3
Brg4   is unmoved

With these adjustments, the catenary of the rotor system is shown in Figure 5.8.

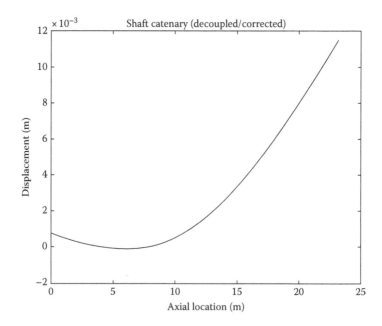

FIGURE 5.8
Coupled rotors with appropriate correction.

Notice that this solution is not unique; having obtained the relative positions of the two rotors, the complete assembly may be rotated. For example, it may be more convenient to move some bearings rather than others. An alternative solution may be reached by raising bearing 4 rather than lowering bearing 3, together with appropriate movements of the other position. In effect, the engineer is free to 'rotate' the arrangement to suit operational needs provided the relative positions and orientations of the rotors are maintained. Whilst the numbers change, the effect is just the same – rotating the orientation of the rotor so as to make the mating faces parallel and concentric. In this example, some negligibly small movements have been considered, but on a large unit, very appreciable offsets arise often with several centimetres difference in height between the HP turbine and alternator. However, as illustrated by comparing Figures 5.7 and 5.8, the alignment adjustment eliminates, or at least significantly reduces, the bending moments in the rotor train.

5.2.4 Misalignment Excitation: A New Model

After mass imbalance, rotor misalignment is the second most prevalent problem in rotating machinery. It is most significant on large multi-bearing machines with two bearings per rotor. Some more recent turbo-generators have only a single bearing per rotor, and these machines are less vulnerable to alignment issues although this type of design does introduce other problems. To a lesser or greater extent, alignment is an issue for all rotating machinery and in view of this, it is surprising that the research findings are rather limited. In marked contrast to the imbalance case, there is not even complete agreement on the symptoms and even more diversity on the explanation. This is probably because there are a number of variables influencing the way in which a rotor will respond to misalignment and the overall analysis of the situation is somewhat complex.

Most authors discuss the generation of harmonics of shaft speed, predominantly the 2X component, yet Patel and Darpe (2009) found in rig trials that 3X was the dominant harmonic. Where flexible couplings are involved, it is easy to appreciate how non-linearities arise from relatively straightforward kinematics, but with rigidly coupled rotors, the situation is much less clear. Al-Hussain and Redmond (2002) developed the equations of motion of two rotors with parallel misalignment, but could not establish any mechanism for directly generating harmonics. The paper suggests that harmonics arises from the non-linearity of the bearings as the rotor is displaced from its normal operating location within the bearing. Other authors have suggested this mechanism, but it is not entirely satisfactory as it is unclear why the harmonics should always arise in this way.

A more direct way to approach the problem is to focus attention on the coupling itself, because this will always be in some sense, the weakest (i.e. most flexible) part of the rotor train.

In a later paper, Redmond (2010) assumes different coupling stiffness values in the two orthogonal direction, rotating with the shaft and he successfully demonstrates that this leads to the generation of harmonics caused by time-varying coefficients rather than non-linearities. This gives some prospect of explaining at least some of the observed behaviours, but like other contributions, it is incomplete in that asymmetries are simply assumed without a deeper explanation. Lees (2007) reached similar conclusions by a rather different route, which is outlined later. That paper assumed that the shaft torque is transmitted by the coupling bolts. This is clearly a rather extreme case but one that helps to clarify the issues. The point to make here is that this model leads to coupling stiffness matrices which are time dependent. These two papers provide some for the development of a more realistic model of (so-called) rigid couplings.

Figure 5.9 shows a simplified solid (rigid) coupling. Two abutting discs are considered with a parallel offset (at this stage, angular offset is ignored although this is readily analysed using the same approach). The two flanges are assumed to be rigid and are connected by a set of N bolts of which only two are shown in the figure for the sake of clarity. Assume that on the flange of the first rotor, the bolt holes are on a circle of radius r, concentric with the shaft. On the second rotor, however, since the shaft is displaced, yet the bolts connect without stress, the centre of the circle of holes must be displaced from the rotor centre by, say, δ. This is the model developed by Lees (2007), but here a rather more complete picture of the motion is needed.

Under normal operation, the N bolts will be tightened to a specified degree, monitored either in terms of torque or bolt extension. Let us assume a bolt torque set at τ and this will ensure a thrust P between the faces. The torque (T) will be transmitted partly by friction and partly via shear of

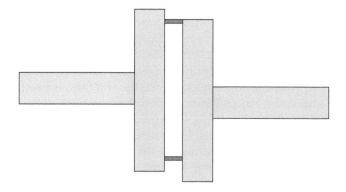

FIGURE 5.9
Simplified rigid coupling. (Reproduced from Lees, A.W., *J. Sound Vibrat.*, 305, 261, 2007. With permission.)

the bolts. The case considered by Lees (2007) represents a limiting case in which the frictional contribution tends to zero. But to take the analysis further, the total stresses arising in the bolts must be considered and these will include terms arising from the bending moments at the coupling due to imbalance excitation. However, there is clearly no direct means of assessing this effect and a model is needed to estimate the bending moments from measured vibration data. Given that at least part of the torque is transmitted by the friction between the abutting flanges and the frictional forces will be substantially effected by moments acting at the coupling, the effective (local) centre of rotation may change. These effects are now considered.

5.2.4.1 Moments due to Imbalance

The imbalance response at speed Ω is given by

$$\mathbf{x} = \left[\mathbf{K} + j\Omega\mathbf{C} - \Omega^2\mathbf{M} \right]^{-1} \mathbf{M}\Omega^2\varepsilon \tag{5.6}$$

where ε is the mass eccentricity vector. Usually, there will only be measurements of the rotor motion available at or close to the bearings, but the model may be used to *recover* estimates of the motion at all the other nodal points, subject of course to all the usual concerns on model accuracy and other uncertainties. Let us say the coupling being studied is at node m of the model, then the four degrees of freedom at the coupling will be $\left[4m-3, 4m-2, 4m-1, 4m \right]$ and to calculate the bending moment, the motion of nodes $m-1$ and $m+1$ must be considered. To derive the bending moment, recall the derivation of the basic beam finite element set out in Chapter 3. For the sake of clarity, the Euler beam formulation is used here. The displacements within the chosen elements are given by

$$u(z_e) = a_0 + a_1 z_e + a_2 z_e^2 + a_3 z_e^3$$
$$v(z_e) = b_0 + b_1 z_e + b_2 z_e^2 + b_3 z_e^3 \tag{5.7}$$

z_e is the axial position within the element. Now the bending moments in the two orthogonal directions are given by

$$M_x = EI\left(\frac{\partial^2 v}{\partial z_e^2} \right) = 2b_2 + 6b_3 z_e \quad M_y = EI\left(\frac{\partial^2 u}{\partial z_e^2} \right) = 2a_2 + 6a_3 z_e \tag{5.8}$$

So the problem now is to determine the parameters a and b and to do this, it is convenient to consider the two directions separately. Taking the first direction, knowing the displacements at the two nodes of the element, the motion can be written as

$$
\begin{Bmatrix} u_{4m-7} \\ u_{4m-4} \\ u_{4m-3} \\ u_{4m} \end{Bmatrix} = \begin{bmatrix} 1 & 0 & 0 & 0 \\ 0 & 1 & 0 & 0 \\ 1 & z_e & z_e^2 & z_e^3 \\ 0 & 1 & 2z_e & 3z_e^2 \end{bmatrix} \begin{Bmatrix} a_0 \\ a_1 \\ a_2 \\ a_3 \end{Bmatrix}
\tag{5.9}
$$

Hence, the solution may be written as

$$
\begin{Bmatrix} a_0 \\ a_1 \\ a_2 \\ a_3 \end{Bmatrix} = \begin{bmatrix} 1 & 0 & 0 & 0 \\ 0 & 1 & 0 & 0 \\ 1 & z_e & z_e^2 & z_e^3 \\ 0 & 1 & 2z_e & 3z_e^2 \end{bmatrix}^{-1} \begin{Bmatrix} u_{4m-7} \\ u_{4m-4} \\ u_{4m-3} \\ u_{4m} \end{Bmatrix}
\tag{5.10}
$$

Similarly, motion in the v direction yields

$$
\begin{Bmatrix} b_0 \\ b_1 \\ b_2 \\ b_3 \end{Bmatrix} = \begin{bmatrix} 1 & 0 & 0 & 0 \\ 0 & 1 & 0 & 0 \\ 1 & z_e & z_e^2 & z_e^3 \\ 0 & 1 & 2z_e & 3z_e^2 \end{bmatrix}^{-1} \begin{Bmatrix} u_{4m-6} \\ u_{4m-5} \\ u_{4m-2} \\ u_{4m-1} \end{Bmatrix}
\tag{5.11}
$$

With these results, the bending moments in the element immediately before the coupling are defined and using the same approach, those in the element after the coupling can also be evaluated. The shear force can also be calculated from these parameters as will be mentioned later in the discussion. Note that since only displacement and slope are continuous across element boundaries, different estimates for bending moments and forces are obtained in the two elements. The optimum estimate is simply the mean of the corresponding values.

Hence, given an imbalance on the rotor, there will be a rotating bending moment vector at the coupling, or to be precise, at the flanges of the coupling.

5.2.4.2 Changing the Centre of Rotation

The bending moment imposes a gradient on the interfacial pressure distribution on the flanges and this has the effect of displacing the centre of rotation by some distance, Δ, whose magnitude depends on both the degree of misalignment and the imbalance excitation. The equation of motion will then become

$$\mathbf{Kx} + \mathbf{C\dot{x}} + \mathbf{M\ddot{x}} = \mathbf{M}\Omega^2\varepsilon + F(\Delta) \tag{5.12}$$

Δ may of course be strongly time dependent, and since it has arisen from the interaction of imbalance and misalignment, it will often have a number of harmonics of the rotation speed Ω and so, the response will contain harmonics of shaft speed. F is the force due to the change in stress field.

A limiting cases can be examined relatively easily: the case considered by Lees (2007) is one in which all of the torque is transmitted through the bolts and this is outlined in Section 5.2.4.3.

To examine a more likely scenario, we must consider the way in which real couplings operate and, in particular, the way in which torque is transmitted. The model of torque transmission via the bolts presented in Section 5.2.4.3, although of some interest insofar as it presents a limiting case, is highly idealised and we now examined a more realistic situation. The coupling faces will be held together with some pressure arising from the bolt tension. Torque is then transmitted partly through shear in the bolts but mainly through friction. The precise balance of these two would require a detailed three-dimensional (3D) finite element study and this is beyond our current scope, but considerable progress can be made with some analytic considerations.

With perfect alignment (and ideal balance), there will be even pressure across the face and the centre of rotation will coincide with the geometric centre. However, the combination of misalignment and imbalance leads to a cyclic bending moment across the face. As illustrated in Figure 5.10, when the bending moment is tending to separate the upper portion of the coupling faces, the centre of rotation will move downwards by a distance Δ. This may be calculated by equating the moments above and below the unknown centre of rotation. Given that, the interfacial pressure may be written as

$$P(y) = P_0\left(1 - \alpha\frac{y}{r}\right) \tag{5.13}$$

Then the task is to establish a relationship between Δ, the displacement of the centre of rotation, and α, the pressure gradient (or equivalently, the resultant bending moment). This relationship can be established from the equation

$$2\int_{-\Delta}^{r} P(y)(y+\Delta)\sqrt{r^2-y^2}\,dy = 2\int_{-\Delta}^{-r} P(y)(y-\Delta)\sqrt{r^2-y^2}\,dy \tag{5.14}$$

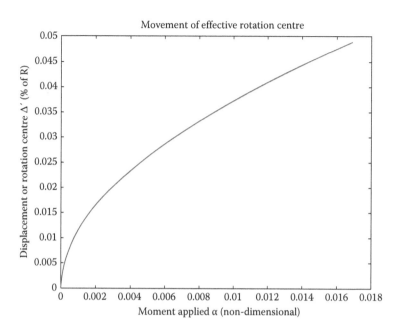

FIGURE 5.10
Change of rotation centre with bending moment.

It is convenient to re-write this in a non-dimensional form by setting $s = y/r$ then Equation 5.14 may be expressed as

$$\int_{-\Delta'}^{1} P_0(1 - \alpha s)r^3(s + \Delta')\sqrt{1 - s^2}\,ds = \int_{-\Delta'}^{-1} P_0(1_0 - \alpha s)r^3(s - \Delta')\sqrt{1 - s^2}\,ds \qquad (5.15)$$

The factor r^3 cancels and $\Delta' = \Delta/r$. The equation may be re-arranged using definite integrals and written as

$$I_1 + \Delta'I_2 - \alpha I_3 - \alpha\Delta'I_1 = I_4 - \Delta'I_5 - \alpha I_6 + \alpha\Delta'I_4 \qquad (5.16)$$

Collecting terms, the values of α corresponding to each displacement Δ' may be established as

$$\alpha = \frac{I_1 + \Delta'I_2 - I_4 + \Delta'I_5}{I_3 + \Delta'I_1 - I_6 + \Delta'I_4} \qquad (5.17)$$

The integrals used in this expression are

$$I_1 = \int_{-\Delta'}^{1} s\sqrt{1-s^2}\,ds = -\left[\frac{(1-s^2)^{3/2}}{3}\right]_{-\Delta'}^{1}$$

$$I_2 = \int_{-\Delta'}^{1} \sqrt{1-s^2}\,ds = \left[\frac{s\sqrt{1-s^2}}{2} + \frac{1}{2}\sin^{-1}s\right]_{-\Delta'}^{1}$$

$$I_3 = \int_{-\Delta'}^{1} s^2\sqrt{1-s^2}\,ds = \left[\frac{s(1-s^2)^{3/2}}{4} + \frac{s\sqrt{1-s^2}}{8} + \frac{\sin^{-1}s}{8}\right]_{-\Delta'}^{1}$$

$$(5.18)$$

$$I_4 = \int_{-\Delta'}^{-1} s\sqrt{1-s^2}\,ds = -\left[\frac{(1-s^2)^{3/2}}{3}\right]_{-\Delta'}^{-1}$$

$$I_5 = \int_{-\Delta'}^{-1} \sqrt{1-s^2}\,ds = \left[\frac{s\sqrt{1-s^2}}{2} + \frac{1}{2}\sin^{-1}s\right]_{-\Delta'}^{-1}$$

$$I_6 = \int_{-\Delta'}^{-1} s^2\sqrt{1-s^2}\,ds = \left[\frac{s(1-s^2)^{3/2}}{4} + \frac{s\sqrt{1-s^2}}{8} + \frac{\sin^{-1}s}{8}\right]_{-\Delta'}^{-1}$$

If, as assumed here the axial torque transmission is through the friction at the couplings, this is limited by the pressure of the interface P_0 and this is set by the tension of the bolts. Assuming that theses bolts have been evenly tightened, the pressure gradient term α is governed by the external bending moments arising from imbalance and misalignment bearing forces. From Equation 5.13, it is clear that α is dimensionless and has scaling such that with a value of unity, the pressure disappears on one side of the coupling face.

The simplest way to solve this equation is to plot the slope (α) for each value of the offset (Δ') and this plot is shown as Figure 5.10.

Actually, this curve is fairly close to a linear relationship and such a simple relationship will suffice for our current investigative study. In a real case, one would anticipate that the movement would be at the lower end of the scale shown, but the an extended scale is retained to clarify the argument.

In summary, the mechanism suggested here is that the combined bending moments due to misalignment and imbalance change the distribution of interfacial friction at the coupling, thereby shifting (locally) the centre of rotation. A complete analysis of this effect would require a 3D model, but insight can be gained from the arguments presented here.

5.2.4.3 Influence of the Bolts

The situation described so far assumes that torque is transmitted entirely by way of the interfacial friction. In all cases, the bolt tightness will set the limits of bending moment that can be sustained, but in general, torque will be transmitted partly by the bolts and partly through friction. In the analysis of the previous section, transmission is entirely through friction, but the other extreme case which can be modelled easily postulates that the torque is transmitted entirely via the coupling bolts. In reality, transmission will be through a combination of these two, but at the time of writing, there is no appropriate formulation of the overall problem.

The case considered comprises two rotors that are not co-axial but have a relative displacement δ. For the sake of simplicity, it is assumed that the first shaft is rigid and has a high moment of inertia, whilst the second has some flexibility. The two rotors are connected by a series of N bolts, which on the first shaft are distributed around the perimeter at some radius r from the axis of the shaft. Initially, it is assumed that the two portions of the assembly rotate at the same speed, but simple geometry dictates that the two pairs of holes cannot remain aligned, since they rotate about different axes.

On the flange of rotor 1, the coupling bolts will be arranged about the centre. Let the number of bolts be N, equally positioned around the circumference of the first rotor. Then at time zero, the position of bolt j is given by

$$x_j = r\cos((j-1)\theta) \qquad \qquad \qquad x_j = r\cos\theta_j$$
$$\qquad\qquad\qquad \text{where } \theta = \frac{2\pi}{N} \text{ or more simply,} \qquad\qquad\qquad (5.19)$$
$$y_j = r\sin((j-1)\theta) \qquad\qquad\qquad\qquad y_j = r\sin\theta_j$$

On the other rotor, the holes are not distributed around the centre, but offset. Simple geometry shows that the position of the jth bolt relative to the centre of this rotor is

$$\begin{Bmatrix} X_j \\ Y_j \end{Bmatrix} = R_j \begin{Bmatrix} \cos\varphi_j \\ \sin\varphi_j \end{Bmatrix} + \begin{bmatrix} \cos\phi & \sin\phi \\ -\sin\phi & \cos\phi \end{bmatrix} \begin{Bmatrix} 0 \\ \delta \end{Bmatrix} \qquad (5.20)$$

where

$$R_j = \sqrt{1 + \delta^2 + 2\times\delta\times\cos(j-1)\theta}$$
$$\qquad\qquad\qquad\qquad\qquad\qquad\qquad\qquad\qquad (5.21)$$
$$\varphi_j = \tan^{-1}\left(\frac{\delta + r\sin((j-1)\theta)}{r\cos((j-1)\theta)}\right)$$

To start with an initially simple model, assume that the two rotors rotate at speeds Ω, $\dot{\phi}$, respectively, where the speed of the first rotor, Ω, is taken to be constant. Then the kinetic energy is

$$T = \frac{1}{2}J_1\Omega^2 + \frac{1}{2}J_2\dot{\phi}^2 \tag{5.22}$$

where J_1 and J_2 are the respective polar moments of inertia of the two rotors (assuming them to be rigid – this is easily generalised once the formulation is completed). The potential energy is

$$U = \frac{K_b}{2}\sum_{j=1}^{N}\left[(r\cos(\theta_j + \Omega t) - R_j\cos(\varphi_j + \phi) - \delta\sin\phi)^2\right]$$

$$+ \frac{K_b}{2}\sum_{j=1}^{N}[(r\sin(\theta_j + \Omega t) - R_j\sin(\varphi_j + \phi) - \delta\cos\phi)^2] \tag{5.23}$$

Taking nominal values for the stiffnesses of the coupling bolts as K_b, the stored energy varies as the rotor system rotates. The motion of the two flanges is illustrated in Figure 5.12 showing the positions of the bolt locations (only three bolts are shown for the sake of clarity). This variation is not sinusoidal and has some harmonic content. Note the locus of the coupling bolts: those on rotor 1 simply follow a single circle, whereas those on rotor 2 each follow a different circle.

The analysis so far has implicitly considered only torsional motion, but in general, we wish to examine a system with lateral motion as well. The full development of the equations of motion can be found in Lees (2007) and the principles are illustrated in Figure 5.11.

We now introduce a simplified model to illustrate the nature of the parametric excitation arising in a misaligned system. The two shafts are connected by a series of N bolts. The first shaft is rigid and has a very large torsional inertia. The each coupling bolt is perfectly positioned at radius r on rotor 1, but on the coupling face of the second rotor, the bolt holes are offset. The second rotor has torsional inertia J_2 and stiffness values k_x, k_y and this rotor has displacements u, v, φ. The arrangement is shown in Figure 5.9.

The analysis of the dynamics is somewhat involved, but after some algebra, Lees (2007) showed that the governing equation is

$$\mathbf{Kx} + \Delta\mathbf{K}(t)\mathbf{x} + \mathbf{C}\dot{\mathbf{x}} + \mathbf{M}\ddot{\mathbf{x}} = \mathbf{F}(t) \tag{5.24}$$

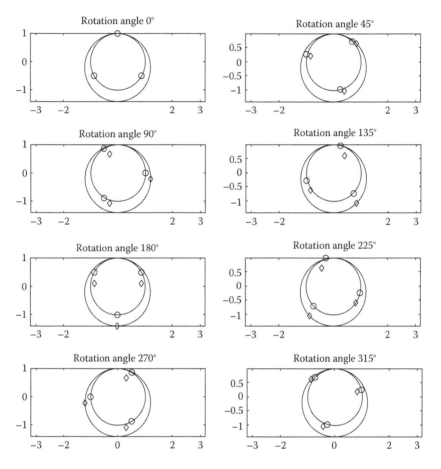

FIGURE 5.11
Motion of adjoining flanges. (Reproduced from Lees, A.W., *J. Sound Vibrat.*, 305, 261, 2007. With permission.)

where

$$\mathbf{F} = \frac{NK_b\delta}{2} \left\{ \begin{array}{c} \cos\Omega t \\ \sin\Omega t \\ \delta \\ -\cos\Omega t \\ -\sin\Omega t \\ -\delta \end{array} \right\} \qquad (5.25)$$

and

$$\Delta K(t) = \frac{NK_b\delta}{2} \begin{bmatrix} 0 & 0 & \sin\Omega t & 0 & 0 & -\sin\Omega t \\ 0 & 0 & \cos\Omega t & 0 & 0 & -\cos\Omega t \\ \sin\Omega t & \cos\Omega t & 0 & -\sin\Omega t & -\cos\Omega t & 0 \\ 0 & 0 & -\sin\Omega t & 0 & 0 & \sin\Omega t \\ 0 & 0 & -\cos\Omega t & 0 & 0 & \cos\Omega t \\ \sin\Omega t & -\cos\Omega t & 0 & \sin\Omega t & \cos\Omega t & 0 \end{bmatrix}$$

(5.26)

N is the number of connecting bolts, bolt stiffness K_b and δ is the offset. Similar expressions can be derived for angular offsets.

The key point to observe in this is that the equation, although linear, has coefficients which vary in time and this will give rise to harmonics of shaft speed in the vibration measurements. Physically this comes from the offset between adjoining shafts, or sections, which means that the transmitted torque on the shaft is converted to a torque plus a resultant force. On some systems, this might simply impose requirements, albeit demanding ones, on machine set-up, but in many cases, the situation is a little more complicated.

5.2.5 Towards an Overall Model

Amidst this fairly extensive mathematical analysis, it is worth reflecting on the physical basis of the phenomenon. Fundamentally, because the centre of rotation and geometric centre do not coincide, an input torque is resolved into a couple and a force. Furthermore, the way in which this resolution is achieved is a function of time. Nevertheless, the discussion is valid in principle for any system in which the transmission has some angular variability and it is difficult to imagine a practical machine in which this is not the case. Of course, the mechanism of coupling between torsional and lateral motion discussed earlier is not the sole source of second harmonic excitation.

Other factors contributing to harmonic excitation in misaligned systems include static change of loading position changes the restoring loads and torque vector rotation. If there is non-linearity in the bolts or bearings or changes of internal clearance leading to variations of pressure or electric field, these influences can also modify the harmonic response. Indeed this is an area in which there are a number of unresolved issues and it is perhaps the multiplicity of sources which have stifled progress on a thorough understanding of misalignment. The confused situation is more pertinent, since misalignment remains one of the most significant practical problems in rotating machinery. In summary, both of the cases for torque transmission have been shown to give rise to vibration components at harmonics of the rotation frequency. There is a need for a comprehensive formulation in order to give quantitative predictions of behaviour in real machines.

5.3 Cracked Rotors

The dynamics of cracked rotors has been an active area of study for a considerable time and this activity is driven by the severe problem of assuring rotor integrity. Over the years, there have been an appreciable number of cracked rotors, most of which have been diagnosed in time to avoid catastrophic consequences. Such failures incur very high costs in terms of capital damage, loss of production and, not least, danger to personnel, and, consequently, it is of utmost importance to recognise incipient faults from operational data. A pre-requisite to such a facility is an understanding of the dynamics of a cracked rotor. This is a topic in which there have been many papers over the last few decades, but real progress in understanding has been somewhat limited in the last 20 years – it would appear that the field is in a phase of diminishing returns.

Rotating machines are often subject to high stresses and on occasions, this has manifested via cracking of the main rotor. Whilst rotor cracking is a relatively rare fault, it is potentially very dangerous presenting a severe economic danger. Because of this, it is important to understand the characteristics of the vibrational behaviour of a cracked rotor. Research carried out in this area over the last 30 years has analysed the effect of a transverse crack, the most common category although torsional loadings give high-stress components at an angle to the axis of the rotor and corresponding crack geometries have been observed.

Let us now consider the dynamics of a rotor with a transverse crack. This is illustrated in Figure 5.12. We assume in this section that the axis of the

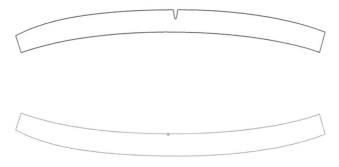

FIGURE 5.12
Alternate opening and closing of a crack.

rotor is horizontal. (In the case of a vertical shaft, the problem of detection is somewhat more complicated.) As the shaft rotates, the stiffness rotor will vary. This is because when the orientation is such that there is a compressive stress across the crack faces (due to the self-weight bending), the crack will be closed and the shaft will have its full stiffness, that is the crack is effectively absent. However, half a revolution later, the crack will be open under the action of tensile stresses. This causes a reduction in the effective stiffness of the rotor. The portion of the crack below the rotor's neutral axis will be open and this degree of opening will determine the dynamic characteristics. The calculation of the dynamics can be significantly simplified by observing that in many instances, the vibration amplitude will be significantly less than the shaft deflection due to gravity (i.e. the so-called self-weight bending). With this assumption, the non-linear equation of motion is converted to a linear equation with time-varying coefficients, making the analysis very much more tractable.

The behaviour of a rotating cracked shaft received extensive discussion from 1975 onwards with notable early contributions from Mayes and Davies (1976, 1984), Davies and Mayes (1984), Gasch (1993) and Henry and Okah-Avae (1976). It was noted that, in the case of a horizontal rotor, the self-weight bending plays a very important role in the dynamics. At any stage in the cycle, the crack may be open or closed – indeed, if operating very close to a critical speed, the crack may remain open throughout a full shaft revolution if the unbalance is very high. However, for modest levels of unbalance and away from critical speeds, the equations governing the shaft motion may be linearised by observing that the bending moment due to self-weight bending will dominate over inertial terms. Therefore, those portions of the crack below the neutral axis at any instant may be considered to be open, whilst those above the axis will remain closed. It is assumed that only the open portions of the crack change the rotor's stiffness and hence, this stiffness will be a function of the orientation of the rotor.

This is also true for the case where the rotor has been taken out of the machine for examination and hung (horizontally) from a cradle. There are two key steps to be established for the understanding of cracked rotors: First, the reduction in stiffness due to a given crack is established and second, the size of effective crack for each orientation is discussed.

Mayes and Davies were among the first authors to address the issue of cracks in real machines and in a series of elegant papers introduced significant progress in the analysis of the topic. Two very significant advances were in the treatment of natural frequencies and the analysis of the forced response.

5.3.1 Change in Natural Frequencies

The study of the change in natural frequency due to a chordal crack stems from basic fracture mechanics. The connection between crack depth and change in the rotor's dynamic properties was derived by considering the change in energy as a static force was applied to a rotor. The situation is shown in Figure 5.12 as discussed below.

This figure shows the shaft in two different orientations. The key point is the role played by gravity, causing the shaft to sag to some extent between the bearings. As the shaft rotates, for the part of the cycle when the crack is in the upper portion of the rotor, bending moments from the weight will tend to hold the crack closed and at this instant, the full stiffness will be retained.

Half a cycle later, the crack will be below the neutral axis of the rotor and the weight-induced bending moments will tend to open the crack, perhaps inducing further growth. In this case, the shaft will be weakened to some extent and it is clear that as the shaft rotates, its stiffness will vary with some cyclic pattern, although at this stage, it is not clear what the precise amplitude or pattern of this variation takes. Although there have been a very large number of research papers on this topic, in truth, the detailed pattern of the variation is not of great importance as shown by Penny and Friswell (2002): every crack in a real machine will have a different size and profile and will give rise to effects which differ in detail. What is crucially important for the practicing engineer is the recognition of the salient effects characteristic of a rotor crack.

In their paper, Mayes and Davies (1976) related the change in natural frequency to the crack depth and location. The analysis considered the energy required to increase the deflection of a crack beam (rotational effects were neglected in this analysis). The energy (equal to the force applied multiplied by the increase in deflection) is accounted for partly by the potential energy stored in the beam and party by the energy required to grow the crack. A key concept here is the strain energy release rate, G, given by

$$G = -\frac{\partial U}{\partial A} \tag{5.27}$$

where
    U is the stored energy
    A is the area of the crack

This can be related to the stress intensity factor by the relationship (for plane stress)

$$G = \frac{K_I^2}{E} \tag{5.28}$$

K_I is the stress intensity factor. The mathematics of the stress intensity factor is somewhat involved, but for the present discussion, it is not necessary to delve into this. A range of expressions are available for different specimen geometries and crack profiles, but the important point to note is that this factor essentially describes the stress distribution around the crack. It is not surprising that this is needed to obtain the energy stored in a cracked rotor and hence the change in natural frequencies.

It was shown that (Mayes and Davies 1976)

$$\Delta\left(\omega_n^2\right) = -4\left(\frac{EI^2}{\pi r^3}\right)(1-v^2)F(\mu)\left(y_N''\left(s_c\right)\right)^2 \tag{5.29}$$

where

$$y_N''(s_c) = \left[\frac{d^2 y_N}{dz^2}\right]_{z=s_c}$$, that is the second derivative of the mode shape at the crack location S_c

F is a function of non-dimensional crack depth – the compliance function

y_N is the Nth mode shape and μ is the non-dimensional crack depth, that is crack depth divided by shaft radius. The argument was developed by considering the energy required to extend the crack by some small extent and comparing this to the energy stored in the deformed beam.

The other parameters in this equation are

E is the Young's modulus

I is the second moment of area

r is the shaft radius

v is Poisson's ratio

In principle, the function $F(\mu)$ can be derived from the appropriate stress concentration factor, if this is known. A different approach was taken in Mayes and Davies (1984) where the authors inferred the values of F from a series of experiments with chordal cracks. It was shown that to a very good approximation, this function is just equal to the fractional change in the second moment of area as measured at the crack face. $F(\mu)$, the compliance function, has two components corresponding to two orthogonal directions in line with and at right angles to the crack front. The compliance function, F, is essentially the second moment of area of the remaining (i.e. un-cracked) portion of the cross section. What gives this function a highly non-linear dependency on non-dimensional crack depth, $F(\mu)$, is the fact that the second moment of area is taken about the new mean plane, not that of the original centreline of the shaft. Indeed, it is the form of this (which is a function of the circular geometry) which determines the sensitivity. For a circular cross-section rotor, a chordal crack will need to have an appreciable depth before it can be detected by means of vibration alone. There have been numerous attempts in the literature to develop methods with increased

FIGURE 5.13
The compliance function.

resolution, with some modest successes, but there is a fundamental limitation that a microscopic crack will have little influence on the rotor-dynamic behaviour. Hence, F is a very non-linear function of μ as illustrated in Figure 5.13.

Having established the change in stiffness and natural frequency arising from a crack, it is straightforward to represent these factors in a finite element model. In Mayes and Davies (1984), it is shown that a crack may be represented by reducing the second moment of area of a single element by ΔI. Using a Rayleigh type of approach, it can be shown that if the second moment of area of the element concerned is I_0, then

$$\frac{\Delta I/I_0}{1-(\Delta I/I_0)} = \frac{r}{L_e}(1-v^2)F(\mu) \tag{5.30}$$

where
$\quad L_e$ is the length of the section with reduced properties
$\quad r$ is the radius of the shaft at that cross section

Note that this parameter is at the discretion of the modeller within a reasonable range, but the choice determines the value of second moment of area. For any given chordal crack, there are two values of F, and two orthogonal directions in the plane of the crack. Hence, the parameters of a representative model are now fully specified.

In the prediction of the dynamic behaviour of a cracked rotor for each orientation of the rotor, a suitable crack model must be derived, and there are several steps to achieve this. For the rotor under non-rotating conditions, it is reasonable to assume that the portion of a chordal crack lying below the neutral axis will be opened under the influence of self-weight bending. The actual portion of the crack opening under these circumstances will have a rather complicated shape having its own stress intensity factor. In general, these factors are not known, but the details are not of great importance. For most purposes, it suffices to base calculations of the effects on total opening with a sinusoidal variation. The details of the opening will influence only the relative amplitude of the higher harmonics (see Penny and Friswell 2002).

At each orientation of the rotor, for a given size of chordal crack, the area of open crack must be calculated. The direction of the weaker axis bisects the angle of the open crack portion. The compliance functions in the two orthogonal directions are then evaluated by assuming equivalence with the chordal crack of equal effective area. It is argued that this relatively simple procedure is sufficient to reflect the main physics of the situation. Whilst it could be argued that the equivalence should be based on second moments rather than area, it must be remembered that we do not have the appropriate stress concentration factors to use and this being the case, it is desirable to keep the calculation as simple as possible. The two extremes of crack open and closed are modelled correctly and to a large extent, precise evaluation of the intermediate values is not necessary. Figure 5.14 shows graphically the open and closed portions for a particular crack depth and orientation.

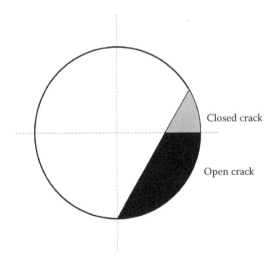

FIGURE 5.14
The equivalent crack. (Reproduced from Lees, A.W. and Friswell, M.I., The vibration signature of chordal cracks in asymmetric rotors, in *19th International Modal Analysis Conference*, Kissimmee, FL, 2001. With permission.)

Example 5.2

As an example, consider the effects of a crack on the natural frequencies of the rotor shown in Figure 5.15. This is a large rotor, similar to a generator rotor with an overall length of 16 m and a mass of 52 t. The bearings are taken as rigid in this example. The diameter of the rotor is 1.2 m over most of the length, but note that in a real machine, this would not all be solid steel and so, the density in the model has been reduced to reflect this.

Chordal cracks are considered with depths of 0.25, 0.5, 0.75 and 1 times the shaft radius and therefore, even the smaller cases are very significant defects. Each of these cracks is considered at three different axial locations, namely, 4, 6 and 8 m from the end. The results were calculated by reducing the stiffness of relevant sections of the rotor in accordance with Equation 5.30 whilst holding the mass constant.

In this study, 16 Timoshenko beam elements were used and in each case, the element of either side of the nominal crack location was reduced. Table 5.2 shows the results of the calculation and a number of important features are apparent. An important first point is that although all the cracks are considered significant, in most cases, the change in natural frequency is modest. For instance a crack of radius depth (i.e. half the cross section) at the 4 m location (or quarter span) reduces the natural frequency by less than ½%. A small change such as this would not be detected on real plant. For each crack depth, the frequency reduction for the first mode increases as the position of the crack is moved towards mid-span.

FIGURE 5.15
Test rotor L = 16 m, D_{max} = 1.2 m, mass 52 t.

TABLE 5.2

Effects of a Crack on Natural Frequencies

Crack Depth (µ)	L/4 (4 m)	3L/8 (6 m)	L/2 (8 m)
Mode 1 (un-cracked 14.61 Hz)			
0.25	14.56	14.52	14.49
0.5	14.46	14.33	14.26
0.75	14.25	13.97	13.82
1	13.61	12.95	12.65
Mode 2 (un-cracked 46.79 Hz)			
0.25	46.51	46.61	46.77
0.5	45.94	46.24	46.72
0.75	44.97	45.63	46.63
1	42.71	44.27	46.42

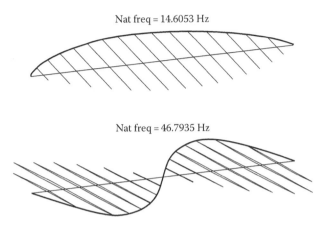

Nat freq = 14.6053 Hz

Nat freq = 46.7935 Hz

FIGURE 5.16
Modes shape of first two modes (in one direction).

This pattern is to be expected from Equation 5.29, giving the important indicator that the effect of a crack on natural frequencies is proportional to the square of the second derivative of the corresponding mode shape. This is physically reasonable as this relates to the bending moment at the location of the defect which will, of course, tend to open the crack.

The logical conclusion is that, for modes which have zero bending moment at a particular location, the introduction of a crack will have no effect. This is clear in the calculated frequencies for mode 2 towards mid-span. The deepest crack at the mid-span shows a frequency change of less than 0.8%. Given the mode shapes shown in Figure 5.16, it is perhaps surprising that there is any change in natural frequency for this mode. The reason for this lies in the modelling: the crack's influence is spread over two elements, or 2 m in this model and so covers regions of the rotor on which the crack has an effect. This highlights the importance of having regard to the potential and limitations of any mathematical idealisation.

Even with very large cracks, the changes in natural frequencies do not provide an adequately sensitive indicator of machine health for use on operational machines, but as discussed in Chapter 9, these small changes can provide a useful tool in crack localisation under laboratory/workshop testing. For initial fault detection, however, the forced response provides a better guide. This is discussed in the next section and attention is focused on machines with horizontal rotors. Those with vertical shafts are a little more complicated to study, because without gravity dominating the crack dynamics, non-linear effects assume greater significance.

In short, forced response is used to detect a crack, and natural frequency changes help to locate it. It should be emphasised that the problem of crack location is not trivial. Only rarely will a crack be obvious to visual inspection without the removal of surrounding components.

5.3.2 Forced Response

Mayes and Davies (1984) observed that for large machines with modest degrees of imbalance, the stress imposed by imbalance is much less than the self-weight bending stresses. This is an important point because it means that the motion can be represented by a linear equation, albeit with time-dependent coefficients, rather than a full non-linear analysis. This makes the study very much more tractable.

The general equation of motion for a rotor may be written, in the usual way as

$$\mathbf{K}x + \mathbf{C}\dot{x} + \mathbf{M}\ddot{x} = \mathbf{F}_{imbalance} + \mathbf{F}_{gravity} \qquad (5.31)$$

The force term due to gravity has been ignored in previous sections, because its effect in other problems is merely to shift the response by a time-independent displacement. However, this is not so with a crack: since the stiffness matrix is time varying, the response at constant speed due to the gravity force is itself time varying, hence the importance of including the gravity term. Before delving into the mathematics of this situation, we can consider the problem physically.

Consider the rotor shown in Figure 5.15. There will, of course, be some vertical deflection, or sag, due to gravity. For a rotor that has no crack, we may deduce the magnitude and shape of this deflection

$$\mathbf{K}\delta = \mathbf{M}\mathbf{S}g \qquad (5.32)$$

where
   g is the acceleration due to gravity
   $\mathbf{S}$ is a selection matrix picking only the vertical components, as used in
      Section 5.2.3

Inverting the stiffness matrix gives

$$\delta = \mathbf{K}^{-1}\mathbf{M}\mathbf{S}g \qquad (5.33)$$

Provided that the stiffness is constant, the deflection in the shaft will also be constant and in fact, the shaft rotates about this 'distorted' axis rather than about the straight line connecting the bearings.

As the crack is at the top of the shaft at an instant in time, there will be compressive stresses in the region of the crack and hence, the crack will have no effect on the catenary. If we now consider the situation after half a revolution of the rotor, then the crack will be at the lower side of the rotor and will be subject to a tensile stress distribution. Hence, the stiffness of the rotor will be reduced by the presence of the rotor. The situation now is as shown in Figure 5.12.

This will result in a greater gravity deflection or 'sag', and hence, the deflection of the rotor will be seen to have a once per revolution deflection, even in the absence of any unbalance forces. Now consider this from a more mathematical viewpoint. Let us first assume that there is no unbalance on the shaft, and therefore, the deflection is only due to gravity. The stiffness will now take the form

$$\mathbf{K} = \mathbf{K}_0 - \mathbf{K}_c(t, u, v) \tag{5.34}$$

where u, v are the deflections. However, in machines, the deflection due to gravity is much larger than the imbalance deflection and therefore, the self-weight bending stresses dictate the dynamics of the crack and consequently, the stiffness can be assumed to be a function only of shaft orientation or equivalently, of time. Following Mayes and Davies (1984), the time dependence is taken (for an ax-symmetric rotor) as

$$\mathbf{K} = \mathbf{K}_0 - \mathbf{K}_c \times (1 + \cos \Omega t)/2 \tag{5.35}$$

This form assumes that the crack is fully open at time zero, but this is easily generalised by the inclusion of a phase parameter. For the present discussion, a phase of zero is assumed. At this stage, the motion can be analysed by considering separately the deflection due to imbalance x, and that due to gravity, δ. With this notation, the equation of motion may be written as (neglecting gyroscopic terms for the sake of simplicity – but their inclusion is straightforward).

The equation of motion is written as

$$\left[\mathbf{K}_0 - \mathbf{K}_c \frac{1 + \cos \Omega t}{2}\right](x + \delta) + \mathbf{C}(\dot{x} + \dot{\delta}) + \mathbf{M}(\ddot{x} + \ddot{\delta}) = \mathbf{M}Sg + \mathbf{M}\Omega^2 \varepsilon \tag{5.36}$$

The stationary part of the solution is given by Equation 5.33.

In Equation 5.36 the stiffness (at constant speed Ω) is a function of time only and although the coefficients are time varying, the equation remains linear. This is important, because it means that the principle of superposition applies (i.e. solutions may be added together). Hence, a solution is possible by summing the terms at different frequencies. Having determined the static response in Equation 5.33, the solution at rotor speed is given by

$$\left[\mathbf{K}_0 - \frac{\mathbf{K}_c}{2}\right](x + \delta_\Omega) + \mathbf{C}(\dot{x} + \dot{\delta}_\Omega) + \mathbf{M}(\ddot{x} + \ddot{\delta}_\Omega) = \mathbf{M}\Omega^2 \varepsilon + \frac{\mathbf{K}_c}{2}\delta_\Omega \tag{5.37}$$

Separating out the part of the solution depending on gravity leads to

$$\left[\mathbf{K}_0 - \frac{\mathbf{K}_c}{2}\right]\delta_\Omega + \mathbf{C}\dot{\delta}_\Omega + \mathbf{M}\ddot{\delta}_\Omega = \frac{\mathbf{K}_c}{2}\delta_0 = \frac{\mathbf{K}_c}{2}\mathbf{K}_0^{-1}\mathbf{M}\mathbf{S}g \qquad (5.38)$$

This gives now the gravity deflection at shaft speed, so the solution is given by

$$\delta_\Omega = \left[\mathbf{K}_0 - \frac{\mathbf{K}_c}{2} - \mathbf{M}\Omega^2 + j\mathbf{C}\Omega\right]^{-1}\frac{\mathbf{K}_c}{2}\mathbf{K}_0^{-1}\mathbf{M}\mathbf{S}g \qquad (5.39)$$

Thus, there will be a once per revolution variation in the rotor position due purely to the effect of gravity and the variation in shaft stiffness due to the crack. This will be of the form $\delta_\Omega e^{j\Omega t}$. This will usually be very small. There will also be higher frequency terms, but these are of less practical importance.

The feature of note is response due to imbalance. From Equation 5.37, the synchronous response to imbalance is given by $\mathrm{Re}\left(\mathbf{x}_0 e^{j\Omega t}\right)$ where

$$\mathbf{x}_0 = \left[\mathbf{K}_0 - \frac{\mathbf{K}_c}{2} - \mathbf{M}\Omega^2 + j\mathbf{C}\Omega\right]^{-1}\mathbf{M}\Omega^2\varepsilon \qquad (5.40)$$

This is just the standard expression for imbalance response. Of greater significance with regard to characteristics of crack rotors is the response at twice the shaft speed. Referring back to Equation 5.37 and extracting second-order terms, the vibration at twice rotation speed is given by $\mathrm{Re}\left(\mathbf{x}_{2\Omega}e^{2j\Omega t}\right)$ where

$$\mathbf{K}_0\mathbf{x}_{2\Omega} + 2j\Omega\mathbf{C}\mathbf{x}_{2\Omega} - 4\mathbf{M}\Omega^2\mathbf{x}_{2\Omega} = \frac{\mathbf{K}_c}{2}(\mathbf{x}_\Omega + \delta_\Omega) \qquad (5.41)$$

This is easily solved to give

$$\mathbf{x}_{2\Omega} = \left[\mathbf{K}_0 - 4\mathbf{M}\Omega^2 + 2j\Omega\mathbf{C}\right]^{-1}\frac{\mathbf{K}_c}{2}(\mathbf{x}_\Omega + \delta_\Omega) \qquad (5.42)$$

almost invariably the second term is negligible, that is $|\mathbf{x}_\Omega| \gg |\delta_\Omega|$.

This analysis has been presented by Sinou and Lees (2005) where a more rigorous discussion is given based on the harmonic balance approach. In that calculation, the non-linear influence of the imbalance deformation was included and the proportion of the crack open was recalculated at each step. The procedure outlined is considered sufficient for most studies.

Equation 5.38 is, of course, central to any study of the dynamics of a cracked rotor, and there are many papers addressing the topic. The second harmonic, evaluated in Equation 5.42, provides an excellent indicator of crack damage. Although there are other, benign, sources of excitation at twice the rotor speed (particularly of electrical machines), the corresponding levels of vibration tend to be very stable. Therefore, any unexplained change in the twice per revolution vibration component (increase or decrease) warrants an urgent investigation. Chapter 9 has a detailed account of the diagnosis of a crack on a turbo-alternator.

5.4 Torsion

The lateral vibrations of a rotating shaft are usually the most demanding in terms of potentially damaging influence on a machine's operation and for this reason, many machines carry the appropriate instrumentation. In any case, these features are readily quantifiable using standard techniques. The same is not true of torsional vibrations which are easily overlooked, yet they can lead to high stresses and major problems in machines. They can be unnoticed as these vibrations occur strictly within the rotor and usually do not transmit to the surrounding structure or components: an exception to this is the case of machines incorporating gearboxes and a case involving this is discussed in Chapter 9.

The fact that these oscillations are not transmitted to the surroundings implies that they are not readily measured, but given appropriate instrumentation, this difficulty can be overcome. A further consequence of considerable concern is that damping of torsional vibration is often restricted to the material damping of the rotor itself and will often be very light. This can be overcome in some systems by the introduction of special flexible couplings which provide damping by either the compression of rubber blocks or the movement of fluids as in the case of hydraulic clutches.

There are two main ways in which torsional vibrations can be measured, but both are rather awkward for different reasons. The first, and perhaps most obvious, of these is the use of a single-toothed wheel on the rotor coupled to a proximal transducer forming a detailed tachometer signal. Because of the high inertia of many machines, the actual variation in angular velocity will be small even with high torque fluctuations, and consequently, considerable signal processing may be required to produce useful, accurate information. However, the effort may lead to invaluable insights into machine operation. An extension to this concept is the use of two (or more)-toothed wheels at different axial positions on the shaft. With careful calibration, such a system can directly measure the twist of a rotor. To put this into context, however, consider the order of magnitude of twist in, say, a 500 MW alternator rotor (being approximately 8 m in length). Assuming that the torque acts across half the length, a straightforward calculation shows that the twist is on order 0.025 radians, corresponding to a displacement at the surface of the rotor of 6.4 mm or a time difference of 82 µs at 3000 rev/min. Note that the fluctuations are likely to be much smaller than this figure and consequently, a combination of rapid data acquisition and effective signal processing is likely to be needed. Chapter 8 shows some appropriate techniques.

A more direct approach can be used to directly measure the transmitted torque on a rotor either be incorporating a torque meter into the machine or fitting a coupling shaft with an appropriate strain gauge bridge arrangement. Such an approach has been used in the investigation of a long-standing problem on a boiler feed pump and was reported by Lees and Haines (1978). Because torque is measured directly, this approach is much less demanding in terms of resolution, but one has the practical problem of retrieving the information. At the time of the investigation (1976), this was transmitted from the shaft by a telemetry system, but a modern alternative would be a shaft-mounted logger for subsequent recovery. In any event, the requirements go well beyond what might be considered standard instrumentation, but the information (and insight) gained is often very valuable. The study is discussed in Chapter 9, but here the underlying principles and methods of analysis are examined.

Torsional excitation is an area where further developments are needed to enhance monitoring capabilities. Usually, at least in the initial stages of an investigation, no quantitative information on the torque variation will be available, but it should not be overlooked as a potential source of problems. So, what can cause variations in transmitted torque? There are a number of possibilities:

a. In fluid-handling machines (including pumps and turbines), the torque generated or absorbed will not be absolutely uniform but will inevitably be composed of a series of pulsations which occur each time an impeller or rotor blade passes a passage in the stator. Hence, a detailed study of the torque profile in time shows a set of patterns, the precise form of which is determined be the detailed hydrodynamic behaviour of the machine. In an ideal system, these pulsations will occur many times per revolution (depending on the number of blades, vanes and stages), but clearly, any departure from uniformity in the components can give rise to excitation at significantly lower frequencies but with increased amplitude.

b. Similarly, in electrical machines, there will always be a small high-frequency variation in torque related to the passage of rotor and stator windings. Once again, any irregularity of the spacings will reduce the frequency but increase the amplitude of variation.

c. Variations can arise from built imperfections or the effects of wear and increase clearances. They can also arise from changes in alignment – one of the many ways in which apparently distinct error mechanisms become related. Indeed, it is this inter-relationships of most of the fault mechanisms which make the ultimate resolution of machine problems such a challenging area of activity.

d. Another source of torsional vibration arises in any system which includes a gearbox. Gears, like any mechanical system, inevitably are imperfect. Gear errors are subject to a series of national and international standards, but the effect of even modest errors can be very significant and the sensitivity of a system to gear errors depends on the characteristics of the complete machine, particularly the torsional critical speeds and their corresponding mode shapes.

The generation of torque fluctuations in gears is basically a displacement-controlled process rather than the more usual situation of force control. This means that the gears must move in such a manner as to accommodate the errors in their profile and twisting moments are generated which ensure that this motion can take place.

The system shown in Figure 5.17 comprises a driver unit (designated by inertia I_1) which drives a single-stage gearbox (inertias I_2, I_3) through a flexible coupling, to a driven machine (inertia I_4) again coupled via a flexible coupling. This is a simple idealisation of a system, yet it retains sufficient detail to be representative of many practical arrangements. It should be noted, however, that this model only represents the torsional motion. In reality, there will be coupling to the lateral motion which will be reflected by vibration monitored on the gearbox bearings. This measurement often provides a good indication of the machine condition although accurate modelling is frequently difficult owing to the difficulty in obtaining an accurate assessment

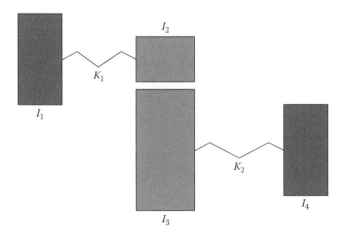

FIGURE 5.17
Idealised torsional model.

of stiffness properties. Concentrating on the torsional motion, the equation of motion may be written as

$$
\begin{bmatrix}
k_1 & -k_1 & 0 & 0 \\
-k_1 & k_1 & 0 & 0 \\
0 & 0 & k_2 & -k_2 \\
0 & 0 & -k_2 & k_2
\end{bmatrix}
\begin{bmatrix}
\theta_1 \\ \theta_2 \\ \theta_3 \\ \theta_4
\end{bmatrix}
+
\begin{bmatrix}
I_1 & 0 & 0 & 0 \\
0 & I_2 & 0 & 0 \\
0 & 0 & I_3 & 0 \\
0 & 0 & 0 & I_4
\end{bmatrix}
\frac{d^2}{dt^2}
\begin{bmatrix}
\theta_1 \\ \theta_2 \\ \theta_3 \\ \theta_4
\end{bmatrix}
=
\begin{Bmatrix}
T \\ P \\ \gamma P \\ D
\end{Bmatrix}
\qquad (5.43)
$$

For the sake of the present discussion, T and D, representing the relevant frequency components of input and output torques, may be taken as steady values as here we wish to focus on the fluctuations introduced by errors in the gear profile. Hence, set $T = D = 0$. P is the transmitted torque arising from errors in the gearbox and γ is the gear ratio.

Although Equation 5.43 shows four angular positions varying; there are in fact only three degrees of freedom, because the two inertias representing the gear pinion and wheel are constrained to remain in contact. This is a reasonable assumption, because if mating gears lose contact during operation, high impact stresses will develop on re-contact, resulting in very rapid wear. In this scenario, the motion is markedly non-linear and detailed analysis becomes very difficult. The objective must be to diagnose incipient problems before this condition is reached. With the assumption of gear contact, the motions of the pinion and wheel are constrained by the equation

$$
\theta_2 + \gamma\theta_3 = \varepsilon_2(\Theta_2) + \gamma\varepsilon_3(\Theta_3) \qquad (5.44)
$$

In this constraint equation, the two error components ε_2, ε_3 are the error profiles of the pinion and wheel, respectively, and these are clearly locked to the two components. Both of these must be cyclic functions of positions around the respective gears. After imposing this constraint, the systems have three degrees of freedom which are chosen to be $\{\theta_1 \quad \theta_2 \quad \theta_4\}^T$. In other words, the dynamics of the system is to be described in a new co-ordinate system given by

$$
\theta = S\theta' + E \qquad (5.45)
$$

where

$$
S =
\begin{bmatrix}
1 & 0 & 0 \\
0 & 1 & 0 \\
0 & -1/\gamma & 0 \\
0 & 0 & 1
\end{bmatrix}
\qquad (5.46)
$$

and the error term is given by $E = \{0 \quad 0 \quad \varepsilon_2(\Theta_2) + \varepsilon_3(\Theta_3) \quad 0\}^T$.

Now substituting (5.45) into (5.44) yields (neglecting damping)

$$\mathbf{KS\theta'} + \mathbf{KE} + \mathbf{MS\ddot{\theta}'} + \mathbf{M\ddot{E}} = \mathbf{F} \tag{5.47}$$

This equation is now pre-multiplied by $\mathbf{S}^T$ to give

$$\mathbf{S}^T\mathbf{KS\theta'} + \mathbf{S}^T\mathbf{KE} + \mathbf{S}^T\mathbf{MS\ddot{\theta}'} + \mathbf{S}^T\mathbf{M\ddot{E}} = \mathbf{S}^T\mathbf{F} = 0 \tag{5.48}$$

Note that to a first approximation, the error function $\mathbf{E}$ is a function of time. In cases of substantial errors, it may become non-linear, since it is dependent on the actual position of the gear teeth, hence including the oscillation. On occasions, this relatively minor effect may become important, giving rise to frequency components in the measured signal which are not related to the rotational speed of the rotor. Of course, even greater non-linear effects are observed if there is a loss of contact in the gears. In either scenario, the basic conclusion would be that torsional oscillations are excessive and some palliative actions would be required.

Flexible couplings are discussed in Section 5.2 in connection with misalignment, but at this point, some general comments on the role of flexible couplings in machinery with regard to torsion are appropriate. There are various types of coupling, some based on Hook's joint principles whilst other types rely on transmission through rubber inserts. Their most common purpose is the reduction of transmitted forces and moments allowing misalignment between driver and driven units which may arise from large differential temperatures or external loads. Ideally, they possess zero lateral and axial stiffness, although this ideal is never fully realised. They do however incorporate considerable torsional stiffness, and these may dominate the dynamics of a small machine such as the model presented in this section. On systems incorporating (torsionally) flexible rotors, the calculation of the stiffness matrix is slightly more involved but easily accomplished using standard FE methods. Flexible couplings of all types inevitably involve some degree of non-linearity in their behaviour.

The dynamics of the geared system is quite complicated and this is reflected in Equation 5.47. It is seen that the motion is driven by $\mathbf{E}$ and its second time derivative. In the model presented, only E_3 is non-zero and this takes the form

$$E_3 = \sum_{k=1}^{n_p} a_k \cos k(\Omega_p t + \theta_2) + \sum_{m=1}^{n_w} b_m \cos m(\Omega_w t + \theta_3) \tag{5.49}$$

where n_p, n_w are the numbers of teeth on the pinion and wheel, respectively, whilst the two rotational speeds are Ω_p and Ω_w in radians/s.

For small oscillations, the non-linear components may be neglected and this just yields sinusoidal variations at the two-shaft speeds and harmonics, the amplitude of which depends on the error profiles on the two gears. However, if wear has become appreciable, and the machine is operating close to a torsional critical, the non-linear effects of Equation 5.49 may become significant. The solution of Equation 5.48 under these circumstances is somewhat involved (not to say tedious!), but the physical form of the result can be readily appreciated. Some analysts in this area represent the forcing term as white noise, but although that way forward is reasonably effective, it is not very illuminating. Observe that because both θ terms have sinusoidal forms, E_3 involves a series of products of sine waves. The net result of this is that the overall excitation involves many frequency components, ranging up to twice the mesh frequency, that is $2n_p\Omega_2/(2\pi)$. The spacing of the components is the reciprocal of the remeshing time, that is the time for a given tooth on the pinion to meet a specified tooth on the wheel. Typically, this time will be given by $T_{rm} = 2\pi n_p/\Omega_2$ assuming n_p, n_w do not have a common factor and this time may be many seconds. Hence, the excitation is a wide-band spectrum whose shape is dependent on the error profiles, expressed in the parameters a_k, b_m.

It is not surprising therefore that the system will select those components of excitation close to a natural frequency. Consequently, in addition to harmonics of shaft speed and harmonics thereof, the system may also generate components of vibration at the natural frequencies, irrespective of the shaft speed (Lees et al. 2011).

The appearance of harmonics is characteristic of either non-linearity or sometimes time-varying stiffness or mass terms. Both of these are highly pertinent to the understanding of rotating machinery and these issues are discussed in the next section. No attempt is made to give a rigorous mathematical analysis, but some discussion is needed to convey physical insight into the phenomena.

5.5 Non-Linearity

To begin this discussion, consider a single degree-of-freedom system described by the equation

$$Kx + C\dot{x} + M\ddot{x} + K_1x^2 = Pe^{jwt} \tag{5.50}$$

If the non-linear term $K_1 = 0$ then the solution for the steady state is easily achieved by assuming a form $x = x_0e^{j\omega t}$. Substituting this into (5.50) gives

$$(K + j\omega C - \omega^2 M)x_0e^{j\omega t} = Pe^{j\omega t} \tag{5.51}$$

The simplification is straightforward, because the first and second derivatives of the function $e^{j\omega t}$ are just constants multiplied by the original function: this basically is the reason why Fourier methods are so useful in the study of vibration problems.

This assumes of course that K, M, C are all constant, varying neither with x nor time. The sinusoidal variations cancel and the expression is easily inverted to give

$$x_0 = \frac{P}{K - \omega^2 M + j\omega C} \tag{5.52}$$

From this, the full solution is easily recovered. Note that a transient solution of Equation 5.50 (with $K_1 = 0$) is only marginally more arduous. The real difficulty arises when non-linearity is present (i.e. $K_1 \neq 0$), because in that case, the equation analogous to (5.51) becomes

$$(K - \omega^2 M + j\omega C)x_0 e^{j\omega t} + K_1 x_0^2 e^{j2\omega t} = P e^{j\omega t} \tag{5.53}$$

Unlike the linear case, it is not possible to cancel the sinusoidal term to simplify the solution procedure. This dilemma is solved by appreciating that the substitution made for x is not adequate for this case as there cannot be a single frequency component involved in the solution. Indeed, this is the important point of this section.

Therefore, a new form of solution is taken as

$$x = \sum_{n=-\infty}^{\infty} a_n e^{jn\omega t} \tag{5.54}$$

Here, of course, the coefficients a_n are not restricted to being real (complexity of the coefficients a_n just corresponds to a shift of phase). The technique for finding a solution comprises substitution of Equation 5.54 into 5.53 and then seeking to match each harmonic term. This procedure is known as the harmonic balance method and is discussed in many text books. It is, however, illuminating to consider the form of the equation with this form of solution substituted. Equation 5.53 becomes

$$\sum_{n=-\infty}^{\infty} (K - n^2 \omega^2 M + jn\omega C)a_n e^{jn\omega t} + K_1 \left(\sum_{n=-\infty}^{\infty} a_n e^{jn\omega t} \right)\left(\sum_{n=-\infty}^{\infty} a_n e^{jn\omega t} \right) = P e^{j\omega t} \tag{5.55}$$

Owing to the double summation, the balancing of the different frequency components is a little involved. The essential point is that the different values a_n

are interdependent and consequently, non-linear systems will generate harmonics of the forcing frequency and this indeed is normally the factor, which reveals the presence of non-linear features in a system. It should be noted that although the system described by Equation 5.50 represents the simplest possible form of non-linearity, the conclusion of the interdependency of various harmonic terms is completely general.

This is similar to the behaviour of gears discussed in the previous section and more fully in Chapter 9. In that case, the non-linearity means that the effective (displacement controlled) forcing involves a product of two series of Fourier terms (in this case, the gear error profiles). This gives rise to excitation at all sum and difference frequencies. The detection of this type of spectrum interaction presents one technique of detecting non-linear behaviour via the so-called bi-spectrum (see Sinha 2006). This is discussed in Chapter 8.

5.6 Interactions and Diagnostics

Inevitably, the various faults discussed in these chapters are discussed individually, but it should be appreciated that many, if not all, of them can be inter-related. This has important practical consequences as it means that it can be difficult to determine the sequence of events giving rise to observed behaviour.

5.6.1 Synchronous Excitation

If the vibration is synchronous with shaft rotation, the root cause could be imbalance or a rotor bend and Chapter 4 discusses the distinction between the two. A bend, however, may present a complicated problem: is it a permanent bend or one brought on by thermal effects. If thermal, then is it the case of a rotor rub (discussed in Chapter 6), or is it caused by irregularities in the fluid flow in the machine? If the cause is rub, then there are a number of possible causes:

a. An initial bend
b. An imbalance
c. Misalignment of bearings
d. Incorrect clearance
e. Thermal bend

It is difficult to give precise rules by which the engineer can distinguish between these possibilities. Often additional data such as bearing oil temperatures will come into consideration, as this can give some insight into bearing loads. Performance data can help give indication of any variation in clearances.

But the four possible causes of the problem are not independent. For instance a rotor may have a small initial imbalance but be in a machine with tight clearance which also suffers some misalignment. It is often these sorts of problems where a number of issues interact and give rise to a fault condition. It may well be that none of the contributory factors, in themselves, would result in unacceptable behaviour.

5.6.2 Twice-per-Revolution Excitation

This is a little less common but still may originate from a combination of root causes. Apart from the possibility of external forcing, the primary reasons for harmonic generation are either some form of non-linear behaviour (discussed in Section 5.5) or some periodic variation in the system parameters. In many respects, the effects of these two are similar, but from an analysis point of view, there is an important distinction: if parameters vary cyclically, but the equation remains linear, the concept of superposition applies and the contribution of different modes may be added. When the equation is truly non-linear, this concept is no longer valid.

The conditions that show harmonic excitation include

   a. Rotor crack
   b. Misalignment
   c. Other types of non-linearity
   d. Electric influence
   e. Rotor asymmetry

The difficulty is that many of these faults are inter-related both within this group and with the factors, giving rise to synchronous vibration. Misalignment and imbalance can both give rise to rotor rub which can in turn lead to a thermal bend and/or the Newkirk effect of a slow cyclic behaviour and it is often difficult to clarify the root cause. Whilst simple rules can help, the only real solution is the development of as complete a picture of the machine's operation as possible. This will include design, build and maintenance practices as well as vibration records.

5.6.3 Asynchronous Vibration

In addition to the components of vibration occurring at shaft speed and its multiples, it is important to give due attention to parts of the vibration signal without a simple relationship to shaft speed. Particular attention is needed to sub-harmonic components below half shaft speed. The principal causes for these problems are

a. External influences
b. Fluid film, or other instabilities
c. Component looseness
d. Rubs

Here again, the categorisation is by no means clear-cut. Rubs can generate asynchronous excitation; indeed, under some circumstances, it can be chaotic, but as discussed, it can also give synchronous and harmonic behaviour. Fluid film instabilities usually arise from misalignment but may also be generated by forces generated elsewhere in the machine. Of the four categories here, the easiest to identify is normally the external influences, simply by applying some limited measurement to neighbouring plant.

With the rather complicated set of inter-relationships between the vibration signals of varying faults, insight into the physical operation of the machine provides the best prospect for problem resolution.

5.7 Closing Remarks

Chapters 4 and 5 together have described some of the main faults observed in rotating machinery, but so far, one important category of fault has not been discussed. This is rub, or to put it more generally, the interaction of a rotor with its stator. This can give rise to a variety of effects and needs some detailed explanation: this is why a complete chapter – Chapter 6 – is assigned to this topic.

Problems

5.1 A machine comprises two identical rotors connected by a rigid coupling. The diameter of the rotor is 200 mm and at the mid-span of each rotor is a disc of mass 250 kg. The rotors are rigidly coupled, and the influence of gyroscopic effects is negligible. Each of the bearings has a constant stiffness of 10^6 N/m and damping of 1000 Ns/m. If the imbalance on disc 1 is 0.004 kg m, and proportional damping can be assumed giving 1% in the first mode, calculate the response at the bearings over the running range of 0–3000 rpm.

5.2 Explain the difficulties of aligning a large multi-bearing machine with flexible rotors.

a. Describe the circumstances under which a machine mounted on oil film bearings can show vibration whose frequency is below half the rotor speed.

b. With the aid of an appropriate diagram, illustrate a method to obtain correct alignment.

c. Two identical rotors are to be connected together. Under the influence of their own weight, the ends of the shaft make an angle of 0.17 min of arc with the vertical. If each end has an overhang of 0.5 m and the spacing between bearings on each shaft is 4 m, what should the alignment be? [*Keep the first rotor fixed.*]

d. If it is now required to keep the two end bearings at the same level as each other, what should be the positions of the other two bearings?

5.3 Assuming that the machine in Question 1 has been arranged with all the bearings in line, determine the bending moment at the coupling when the machine is stationary. With the rotors decoupled, determine the catenaries of the two separate rotors.

5.4 A motor whose rotor has a moment of inertia of 200 kg m^2 is connected through a flexible coupling of stiffness 3×10^6 Nm/rad to a gearbox which is then connected via a coupling of stiffness 8×10^5 Nm/rad. to a pump with a rotor with a moment of inertia 150 kg m^2. The inertias of the pinion and wheel are 20 and 300 kg m^2, respectively, and the gear ratio is 2. Determine the torsional natural frequencies. If there is a cyclic error on the wheel of 10 μm, what is the magnitude of torque oscillations when the input shaft is rotating at 3000 rpm.

5.5 A uniform rotor 8 m in length is supported on rigid short bearings at either end. A 400 kg disc is mounted at mid-span. Determine the change of the first natural frequency due to a chordal crack, 2 m from end 1 with depth of 0.25, 0.5, 0.75 and 1 times the radius.

a. Describe the symptoms of imbalance, bent rotors, rubbing, shaft cracking and half-speed whirl. In each case, indicate the main features of the response in terms of frequency and, where relevant, time. Also indicate any possible methods of rectification. Explain the factors which determine the influence a rotor crack will have on a given mode of vibration.

b. A generator is known to have critical speeds at 600 and 2200 rpm, and this shows on the once per revolution run-down curve. However, over a period of 24 h, the signal at twice running speed gradually increases and a second run-down shows that the twice per revolution component has a small peak at 1100 rpm, together with a very much larger increase at 300 rpm. What can you conclude about the nature and location of the fault? Give a reasoned argument for your conclusions. (Assume the rotor can be modelled as a uniform beam on rigid supports.)

References

Al-Hussain, K.M. and Redmond, I., 2002, Dynamic response of two rotors connected by rigid mechanical coupling with parallel misalignment, *Journal of Sound and Vibration*, 249, 483–498.

Davies, W.G.R. and Mayes, I.W., 1984, The vibrational behaviour of a multi-shaft, multi-bearing system in the presence of a propagating transverse crack, *Journal of Vibration, Acoustics, Stress and Reliability in Design*, 106, 146–153.

Gasch, R., 1993, A survey of the dynamic behaviour of a simple rotating shaft with a transverse crack, *Journal of Sound and Vibration*, 160, 313–332.

Henry, T.A. and Okah-Avae, B.E., 1976, Vibrations in cracked shafts, in *Institution of Mechanical Engineers Conference on Vibrations in Rotating Machinery*, Cambridge, U.K., pp. 15–19.

Lees, A.W., 2007, Misalignment in rigidly coupled rotors, *Journal of Sound and Vibration*, 305, 261–371.

Lees, A.W. and Friswell, M.I., 2001, The vibration signature of chordal cracks in asymmetric rotors, in *19th International Modal Analysis Conference*, Kissimmee, FL.

Lees, A.W., Friswell, M.I. and Litak, G., 2011, Torsional vibration of machines with gear errors, in *9th International Conference on Damage Assessment of Structures (DAMAS)*, University of Oxford, St Anne's College, Oxford, U.K.

Lees, A.W. and Haines, K.A., 1978, Torsional vibrations of a boiler feed pump, *Transactions of American Society of Mechanical Engineers*, 77-DET-28.

Mayes, I.W. and Davies, W.G.R., 1976, The behaviour of a rotating shaft system containing a transverse crack, in *Institution of Mechanical Engineers Conference on Vibrations in Rotating Machinery*, Cambridge, U.K., pp. 53–64.

Mayes, I.W. and Davies, W.G.R., 1984, Analysis of the response of a multirotor-bearing system containing a transverse crack, *Journal of Vibration, Acoustics, Stress and Reliability in Design*, 106, 139–145.

Muszynska, A., 2005, *Rotordynamics*, CRC Press, Taylor & Francis, Boca Raton, FL.

Patel, T.J. and Darpe, A.K., 2009, Vibration response of misaligned rotors, *Journal of Sound and Vibration*, 325, 609–628.

Penny, J.E.T. and Friswell, M.I., 2002, Simplified model of rotor cracks, in *ISMA Conference on Noise & Vibration 27*, Leuven, Belgium, pp. 607–615.

Redmond, I., 2010, Study of a misaligned flexibly coupled shaft system having nonlinear bearings and cyclic coupling stiffness—Theoretical model and analysis, *Journal of Sound and Vibration*, 329, 700–720.

Sinha, J.K., 2006, Bi-spectrum for identifying crack and misalignment in shaft of a rotating machine, *Smart Structures and Systems*, 2, 47–60.

Sinha, J.K., Lees, A.W. and Friswell, M.I., 2004, Estimating unbalance and misalignment of a flexible rotating machine from a single run-down, *Journal of Sound and Vibration*, 272(3–5), 967–989.

Sinou, J.J. and Lees, A.W., 2005, The influence of cracks in rotating shafts, *Journal of Sound and Vibration*, 285, 1015–1037.

6

Rotor–Stator Interaction

6.1 Introduction

Up to this point, rotating shafts have been considered in isolation, somehow remote from their surroundings. Whilst this is a necessary starting point in consideration of such systems, it neglects important facets of machine behaviour. In many instances, the rotor is surrounded by a working fluid and the conditions within and around these fluid regions are those that determine the operational effectiveness of the machine. In this sense, the study of rotor dynamics is an enabling technology that paves the way for effective machine operation: it is not an end in itself.

Whilst the rotor is at the heart of any rotating machine, useful work is only provided through the interaction of the rotating component with its stator, and hence the interaction between the two is of crucial significance. Chapters 4 and 5 discuss a range of common faults related to the rotor, but in this chapter, the scope is widened to consider the rotor's relationship with the stator. There are several ways in which the interaction between stator and rotor can be categorised, but a useful division is as follows:

1. Interaction through the bearings – This has been discussed to some extent in Chapter 4, but here further details are outlined as to the performance of oil film bearings in particular.
2. The interaction via the working fluid – Both through internal channels that exert force and moments on the rotor and fluid-filled bushes and seals which act as subsidiary bearings.
3. Direct interaction with stator through contact – This may take on one of two forms:
   a. Collision and recoil, involving impulsive forces
   b. Slow rubbing leading to heating

A study of these factors is essential in understanding the operation of machines, and it is crucial in recognising a number of fault conditions. Overall, this is an area of considerable complexity and only an overview can be given here.

6.2 Interaction through Bearings

6.2.1 Oil Journal Bearings

Whilst many smaller machines are supported on rolling-element bearings, oil journal bearings (in a range of subtle variations) predominate on large machines. Their predominance is primarily due to their high load-bearing capacity. The analysis of these bearings is somewhat involved and in general requires a numerical solution of Reynolds equation. An added difficulty is that often knowledge of the system parameters, such as clearance and surface condition, is limited. For a given bearing geometry, three parameters (or rather, sets of parameters) are essential to characterise the influence on a machine's behaviour, namely, the stiffness, damping and load capacity. Of these, the stiffness and damping terms are matrices which are functions of shaft rotation speed.

Nevertheless, closed-form solution can be established for short bearings (i.e. $L \ll D$, where L is the axial length of the bearing and D its diameter), as in this case the assumption can be made that the variations in pressure are much greater in the axial direction as compared to the circumferential. Hamrock et al. (2004) give a full analysis of this and other cases, and a more general discussion is given by Friswell et al. (2010). These analytic descriptions do make some important assumptions, namely, that the process is isothermal, the flow is laminar and there is no cavitation within the bearing. A number of books give the analysis of short bearings, but it is worth quoting the results here for the sake of illustration. Childs (1993) shows that the radial and tangential forces are given by

$$f_r = -\frac{D\Omega\eta L^3 \varepsilon^2}{2c^2(1-\varepsilon^2)^2} \quad f_t = -\frac{\pi D\Omega\eta L^3 \varepsilon}{2c^2(1-\varepsilon^2)^{3/2}} \tag{6.1}$$

where
  D is the journal (shaft) diameter
  Ω is the rotational speed (rad/s)
  η is the oil viscosity
  L is the (axial) length of the bearing
  c is the radial clearance
  ε is the shaft eccentricity as a proportion of c

From the relations in Equation 6.1, it is clear that the modulus of f is given by

$$|f| = \sqrt{f_r^2 + f_t^2} = -\frac{\pi D\Omega\eta L^3 \varepsilon}{8c^2(1-\varepsilon^2)^2}\left(\left(\frac{16}{\pi}-1\right)\varepsilon^2+1\right)^{1/2} \qquad (6.2)$$

It is convenient to plot the normalised force f_N such that $|f| = (\pi D\Omega\eta L^3/8c^2)f_N$. Figure 6.1 shows the variation of force with eccentricity, clearly a non-linear relationship.

The derivative of the force (i.e. the stiffness) is also non-linear, and this has major importance. Therefore, given a load on the bearing, it is important to establish the equilibrium running position of the shaft within the bearing. After a little manipulation, it can be shown that if a vertical force f is applied to the bearing, the shaft will have an eccentricity ε which satisfies the equation

$$\varepsilon^8 - 4\varepsilon^6 + \left(6 - S_s^2(16-\pi^2)\right)\varepsilon^4 - \left(4+\pi^2 S_s^2\right)\varepsilon^2 + 1 = 0 \qquad (6.3)$$

where S_s is the modified Sommerfeld number (or the Ocvirk number) and is given by $S_s = D\Omega\eta L^3/8fc^2$. Although there are eight solutions to this equation, only one of these will be between zero and one, and this solution gives the shaft eccentricity relative to the clearance.

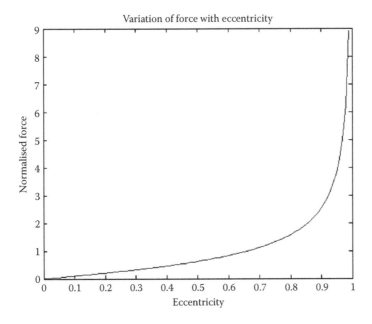

FIGURE 6.1
Normalised force variation.

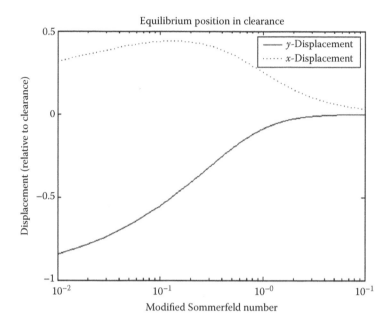

FIGURE 6.2
Variation of position with modified Sommerfeld number.

Figure 6.2 shows the variation of the equilibrium position of the shaft within the bearing assuming that the load is vertical the y-direction as shown in the figure. (Note that this is usually, but not always, the case.)

Having established the running position of the shaft within the bearing, the effective stiffness and damping for small oscillations about this point are established, and these are given as follows:

$$\mathbf{K}_e = \frac{f}{c}\begin{bmatrix} a_{uu} & a_{uv} \\ a_{vu} & a_{vv} \end{bmatrix} \quad \mathbf{C}_e = \frac{f}{c\Omega}\begin{bmatrix} b_{uu} & b_{uv} \\ b_{vu} & b_{vv} \end{bmatrix} \tag{6.4}$$

where

$$a_{uu} = h_0 \times 4(\pi^2(2-\varepsilon^2)+16\varepsilon^2)$$

$$a_{uv} = h_0 \times \frac{\pi(\pi^2(1-\varepsilon^2)^2-16\varepsilon^4)}{\varepsilon\sqrt{1-\varepsilon^2}}$$

$$a_{vu} = -h_0 \times \frac{\pi(\pi^2(1-\varepsilon^2)(1+2\varepsilon^2)+32\varepsilon^2(1+\varepsilon^2))}{\varepsilon\sqrt{1-\varepsilon^2}}$$

$$a_{vv} = h_0 \times 4 \left(\pi^2 (1+2\varepsilon^2) + \frac{32\varepsilon^2 (1+\varepsilon^2)}{1-\varepsilon^2} \right)$$

$$b_{uu} = h_0 \times \frac{2\pi\sqrt{1-\varepsilon^2} \, (\pi^2 (1+2\varepsilon^2) - 16\varepsilon^2)}{\varepsilon}$$

$$b_{uv} = b_{vu} = -h_0 \times 8(\pi^2 (1+2\varepsilon^2) - 16\varepsilon^2)$$

$$b_{vv} = h_0 \times \frac{2\pi(\pi^2 (1-\varepsilon^2)^2 + 48\varepsilon^2)}{\varepsilon\sqrt{1-\varepsilon^2}}$$

and

$$h_0 = \frac{1}{(\pi^2 (1-\varepsilon^2) + 16\varepsilon^2)^{3/2}} \tag{6.5}$$

These rather complicated expressions are sufficient to give insight into the behaviour of hydrodynamic bearings. At zero speed, the shaft will be in contact with the housing and gradually move away as pressure is developed in the oil film. It will be clear that these expressions for stiffness and damping are highly non-linear, and hence any calculated stiffness and damping terms are valid only for small vibrations about the equilibrium position. Any motion in which the amplitude with the journal is significant relative to the bearing clearance will become non-linear, and it is important that such motion is recognised. It is interesting to note the variation in stiffness for different positions within the clearance, and Figure 6.3 shows the variation of bearing stiffness with vertical position (all displacements are considered relative to the radial clearance).

The stiffness is expressed in relative terms, but it is apparent how the stiffness increases dramatically at high eccentricity values.

The study presented here considers plain journal bearings, but there is a range of closely related types of bearings for which similar methods may be applied but clearly with some different results. Grooving of the bearings produces some differences in pressure profiles, and such cases are best analysed numerically with suitable software. Similarly, a fairly common variation is the inclusion of tilting pads to support the oil film. These pads tend to reduce the bearing asymmetry and help with stability considerations.

6.2.2 Rolling-Element Bearings

Whilst hydrodynamic journal bearings predominate in large machines, other types of bearings are important. Many small- and medium-size machines

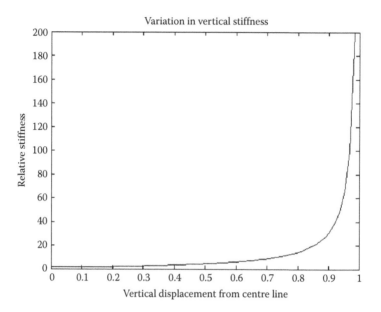

FIGURE 6.3
Bearing stiffness in the vertical direction.

are supported on rolling-element bearings, and these are generally much stiffer than hydrodynamic bearings. The deflection of the balls or rollers can be calculated from Hertzian contact principles. From Kramer (1993), the effective stiffness of ball bearing is

$$k_{vv} = k_b n_b^{2/3} d^{1/3} f_s^{1/3} \cos^{5/3} \alpha \qquad (6.6)$$

where n_b is the number of balls of diameter d, α is the contact angle within the race, f_s is the vertical force and the constant $k_b = 13 \times 10^6 \ \mathrm{N}^{2/3} \ \mathrm{m}^{-4/3}$.

For a roller bearing, the expression is

$$k_{vv} = k_r n_r^{0.9} l^{0.8} f_s^{0.1} \cos^{1.9} \alpha \qquad (6.7)$$

Here there are n_r rollers of length l, and the constant $k_r = 1 \times 10^9 \mathrm{N}^{0.9} \ \mathrm{m}^{-1.8}$.

Equations 6.6 and 6.7 give the stiffness in the vertical direction, and Kramer (1993) gives the ratio of horizontal to vertical stiffness (assuming the load is vertical) as follows:

Ball bearing: $k_{uu}/k_{vv} = 0.46, 0.64$ and 0.73 for 8, 12 and 16 balls, respectively

Roller bearing: $k_{uu}/k_{vv} = 0.49, 0.66$ and 0.74 for 8, 12 and 16 rollers, respectively

In practice, these bearings are often much stiffer than either the rotors or the supporting structures and may be taken as rigid for the purposes of modelling. Indeed it is often preferable to take this approach to avoid numerical instabilities arising from the very high stiffness values.

The essential point with rolling-element bearings is that they give rise to different frequencies in the vibration spectrum. Two frequencies of significance correspond to the rotation of the cage (or train) and that of the balls (or rollers). The angular velocity of the cage is given by

$$\Omega_c = \frac{\Omega}{2}\left\{1 - \frac{d}{D}\cos\alpha\right\} \tag{6.8}$$

and that of the balls by

$$\Omega_r = \frac{\Omega}{2}\left(\frac{D}{d}\right)\left\{1 - \left(\frac{d}{D}\right)^2 \cos^2\alpha\right\} \tag{6.9}$$

where
  D is the pitch diameter of the rotor (or inner raceway)
  d is the diameter of the balls (or rollers)
  α is the angle of contact

Two further frequencies can be generated if there is a defect in the inner or outer raceways; these are

$$\Omega_i = n_b \frac{\Omega}{2}\left\{1 + \frac{d}{D}\cos\alpha\right\} \quad \Omega_o = n_b \frac{\Omega}{2}\left\{1 - \frac{d}{D}\cos\alpha\right\} \tag{6.10}$$

The relative amplitudes of these four frequency components provide a good indication as to the condition of rolling-element bearings. These form the major bearing types on a wide range of light and low-speed machinery.

6.2.3 Other Types of Bearings

6.2.3.1 Active Magnetic Bearings

Whilst the bearings described in the previous two sections predominate, there are other types in operation. Of these, perhaps the most important are magnetic bearings, of which there are two main types: passive and active. In both cases, the major advantage is that by using a magnetic field to levitate and locate the rotor, all contact with the rotor is avoided and frictional losses are extremely low. In the passive type, permanent magnets are used to impose a fixed field on the rotor.

Of somewhat greater interest are active magnetic bearings. In these bearings, the magnetic field is governed by a control system using active feedback operating on shaft position data. They provide negligible drag and are suitable, therefore, for high-speed machines. Costs are high, but for a growing number of applications, they provide an excellent solution to a number of practical problems. The use of active magnetic bearings has been reviewed by Schweitzer (2005).

6.2.3.2 Externally Pressurised Bearings

The externally pressurised or hydrostatic bearing is another type of oil journal bearing. The rotor is supported by an oil film, but the pressure field within the oil is generated by an external pump rather than the rotation of the shaft itself, and this has both advantages and disadvantages. The benefits are that load can be taken at low speed and that operational parameters can be varied externally through the external pump. Any means of applying external control to a rotor system is interesting in the context of developing autonomous machines, and this topic is discussed in Chapter 10. The penalty paid for this facility is additional complication and cost in the oil supply pump.

6.2.3.3 Foil Bearings

Foil bearings have been known for a considerable time, but interest has increased in recent years as part of general attraction towards oil-free systems. Foil bearings are basically air bearings but with the addition of a flexible foil or membrane which takes the load at low speeds. As speed increases the film of air, the pressure field of which this develops viscous and inertia forces arising from the rotation supports load. The main areas of application are light, high-speed rotors.

6.3 Interaction via Working Fluid

6.3.1 Pump Bushes and Seals

As an introduction to this section, let us consider the specific example of a large centrifugal pump. This is a good example to consider, because the fluid has a high density and therefore has a considerable influence on the dynamics of the rotor.

The purpose of a boiler feed pump in a power plant is to raise the pressure of a large quantity of water. This is achieved by passing the water through a series of impellers that give the fluid kinetic energy and at the outer edge of the impeller, the fluid passed into the diffuser section of the stator. There may be several impellers within a pump, but the trend is towards reducing this number with the most recent feed pumps having only two stages.

Each impeller has a number of channels divided by vanes (often six) through which the fluid passes.

The diffuser of each stage comprises of a set of divergent channels, and hence as the fluid passes through, its velocity reduces and, consequently, the kinetic energy is converted to potential energy which in this case is manifested as pressure. This high-pressure fluid is then directed to the entry of the next stage. The number of diffuser vanes is normally one greater than the number of impeller vanes in order to even out the wearing process. In addition to the impellers, there are two neck rings at each stage: these are narrow clearances through which there is a small leakage path for the fluid separating neighbouring pump stages. Some clearance is essential for the smooth operation of the unit, but if it becomes excessive, the efficiency of the pump is impaired: large clearances can also markedly increase the axial thrust on the pump rotor, but a discussion of this is beyond the scope of this text. Other components on the pump rotor are the thrust balance device, the seals (preventing leakage between pump and casing) and the bearings. All of these items exert forces on the rotor and thereby modify the dynamic properties of the machine, often quite dramatically.

Figure 6.4 shows the internals of a modern boiler feed pump. This is a two-stage machine (i.e. there are two impellers) and it operates at a high speed of about 7000 rpm. Over recent years, there has been a marked trend towards higher speeds and fewer pump stages, but many four-stage units remain in service.

Before proceeding, it is worth examining the pressure differential acting across the seals as this is in itself quite a complicated issue. In essence, within a pump, there are two fine clearances that isolate one stage from its neighbour and they act as 'mini-bearings'. As mentioned earlier, their stiffness, damping and inertial characteristics are strongly dependent on the pressure differential across the seal, but this, in turn, is a rather complicated function of the running conditions, and it is worth examining the fundamental issues underlying this situation. In Figure 6.4 the area shaded in grey shows the main flow path the pump on one side of the machine.

The objective here is to obtain some understanding of the forces around the impeller of a centrifugal pump. There are two aspects to this task: our first objective is to establish the pressure differentials across the two seals (denoted A, B and C in Figure 6.5), and then we consider the cyclic forces which arise. To be a little more specific, the objective is to consider the parameters influencing behaviour rather than a precise evaluation, since, as we shall shortly observe, a detailed calculation is difficult and of limited value, since any wear leading to clearance changes will be observed in marked performance changes. Nevertheless, it is important to understand the trends.

Water from the entry (E) flows through the first impeller and most of it will enter the diffuser section (D). This forms part of the stator and comprises divergent channels through which the water slows and consequently increases pressure. From D, however, a small proportion of the fluid leaks back

FIGURE 6.4

Internal components of a boiler feed pump. (Adapted from the Weir Group plc., Glasgow, Scotland, U.K. With permission.)

FIGURE 6.5
A single pump stage. (Reproduced from the Weir Group plc., Glasgow, Scotland, U.K. With permission.)

to E via the neck-ring A. The flow then proceeds through the diffuser channels to be delivered to the inlet of the second impeller at F. At this point, there is another leakage route via the seal at B through which a small flow will return to point D. Similarly, some of the flow at the outlet of the second impeller can return to F via the neck-ring C. The balance drum, on the right-hand side of the figure, also has a significant influence, but we shall not consider this directly.

The task now is to consider how these neck-rings and seals influence the dynamic behaviour.

With zero leakage flow, the pressure at D, P_D is readily calculated by considering a mass of water swirling at half the angular velocity of the rotor. Hence

$$P_D - P_B = \frac{1}{2}\rho\left(\frac{\omega}{2}\right)^2\left(r_3^2 - r_2^2\right) \qquad (6.11)$$

where r_2 and r_3 are the radii at D and B respectively and r is the water density.

And the additional pressure drop with a leakage flow is given by

$$\Delta P = \frac{\lambda}{d^2}Q^2\left(\frac{1}{r_2} - \frac{1}{r_3}\right) \qquad (6.12)$$

where d is the axial separation between the stator and the impeller and Q is the leakage flow rate and λ is the friction factor. This illustrates some of the complexity in an accurate assessment of the pressure distribution within the unit: the important thing is to gain an understanding of the manner in which the parameters vary.

The second leakage path is through the clearance A into chamber E where at entry to the clearance, the fluid swirls at half the shaft speed. By a very

similar argument, the differential pressure across B and C may be calculated. Having established both these differentials, the characteristics of the
sealing clearances can be calculated by an approach first presented by Black
(1969). The calculation proceeds by considering Bernoulli's equation within
the seal, integrating around the circumference and then transforming from a
frame of reference, which rotates with the bulk of the fluid. The result is the
evaluation of forces on the rotor given by

$$F_y = -\varepsilon\left[\left(\mu_0 - \frac{1}{4}\mu_2\omega^2T^2\right) + \mu_1TD + \mu_2T^2D^2\right]y - \varepsilon\omega\left[\frac{1}{2}\mu_1T + \mu_2T^2D\right]z \quad (6.13)$$

$$F_z = -\varepsilon\left[\left(\mu_0 - \frac{1}{4}\mu_2\omega^2T^2\right) + \mu_1TD + \mu_2T^2D^2\right]z - \varepsilon\omega\left[\frac{1}{2}\mu_1T + \mu_2T^2D\right]y \quad (6.14)$$

where D is the differential time operator, $T = L/V$ or the time taken for fluid
to traverse the seal and

$$\mu_0 = \frac{9\sigma}{1.5 + 2\sigma} \quad (6.15)$$

$$\mu_1 = \frac{(3 + 2\sigma)^2(1.5 + 2\sigma) - 9\sigma}{(1.5 + 2\sigma)^2} \quad (6.16)$$

$$\mu_2 = \frac{19\sigma + 18\sigma^2 + 8\sigma^3}{(1.5 + 2\sigma)^3} \quad (6.17)$$

$$\sigma = \lambda\frac{L}{y_0} \quad (6.18)$$

Finally

$$\varepsilon = \frac{\pi}{6\lambda}\left[\frac{\sigma}{1.5 + 2\sigma}\right]R \cdot \Delta P \quad (6.19)$$

6.3.1.1 Influence of System Pressure Distribution

The pressure distribution within the machine is complicated, and it is not
a simple function of shaft speed. To gain some understanding of the com-
plexities involved, we must diverge to consider the pressure developed by
a centrifugal pump. The centrifugal pump is used as an example, but it is

representative of a wide range of fluid-handling machines. Consider a pump comprising N stages, each of which has an impeller radius r and rotating at speed Ω rad/s. Then, the maximum pressure which can be developed by such a pump is given by

$$P_{\max} = \frac{1}{2} N \rho r^2 \Omega^2 \tag{6.20}$$

This, however, is only the beginning of the story, and this pressure represents only a limiting value at zero flow. Any flow through the system will give rise to some dissipative drag terms, partly due to viscous wall effects, turbulent losses and blade/diffuser entry losses, all of which can be represented by linear and quadratic terms in Q, the volumetric flow rate. Taking this into account, we arrive at an expression for the pressure developed as

$$P(\Omega, Q) = P_0(\Omega) + \alpha Q - \beta Q^2 \tag{6.21}$$

Note from this expression that the maximum pressure does not necessarily occur at zero flow. The function shown in Equation 6.21 is known as the characteristic of the pump and this is key information in its operation. The characteristic in general comprises a family of curves $P(Q)$ for different values of speed. It is very desirable that α is negative: this is usually, but not always, the case.

This may be used to establish the conditions under which the machine is operating and thereby evaluate the pressure differentials within the machine. Whilst it may not always be necessary to evaluate detailed numerical values, it is frequently important to have an appreciation of the factors influencing both performance and vibration behaviour in order to optimise plant operation. To achieve this, the properties of the system which is being fed by the pump must be considered. Let this system characteristic be represented by another curve, say P_{sys} then the operating point is determined by the intersection of these two curves as shown in Figure 6.6. P_{sys} will, of course, not be unique but will depend on the settings of various valves within the circuit. The figure shows the operating point on the P, Q plot which becomes purely a function of rotor speed, albeit a fairly complicated one.

The pump characteristic shown in Figure 6.6 can be very helpful in analysing the behaviour of a pump; by predicting the pressure developed, the user is able to make reasonable estimates of the pressure profile within the unit, and comparing measured data with model predictions can offer considerable insight into condition. Some of the work is somewhat speculative, but the insight gleaned into the internal workings of the pump can be very informative.

To clarify this, consider the characteristic plot in a bit more detail. The declining curves show the pressure developed by the pump as a function of volumetric flow rate for different rotational speeds. The variation occurs

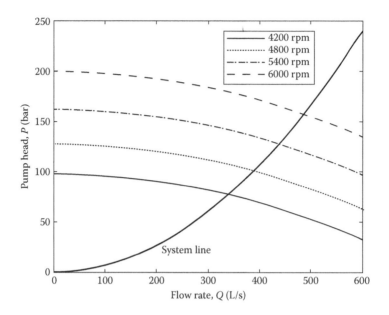

FIGURE 6.6
Typical characteristic curve.

because the fixed blade angles of the diffuser and impeller mean that the flow path will be optimised only for a single flow rate and rotational speed or, at best, only over a narrow range. In practice, machines will be required to operate rather more flexibly and impaired performance is the price paid for this. The plot shows a set of declining curves corresponding to different rotational speeds.

This leads to some intriguing possibilities: Clearly factors which modify the performance of a machine also influence the vibration characteristics. The obvious question arises as to how to correlate changes in performance and vibration measurements to yield added insight into the internal condition of a machine. The complete realisation of this potential is somewhat beyond current capabilities of technology, but research continues to extend the available methods, some of the relevant methods are discussed in Chapter 8.

6.3.1.2 Example of an Idealised System

The relationships expressed in these equations are quite complicated functions of geometry and differential pressure and the variations are now explored using the model shown in Figure 6.7. This is not a representation of any particular machine, but rather a convenient model to illustrate effects of fluid-filled clearances.

The rotor is 2 m in length and supported at each end on bearings with constant stiffness of 5 MN/m. The shaft diameter is 0.15 m, and there are

FIGURE 6.7
Simplified rotor model.

two discs 0.4 m diameter and 0.02 m thick mounted on either side of the mid-span point. At the mid-span, there is a clearance between rotor and stator which is 0.1 m in (axial) length with a radial clearance of 200 μm across which a variable pressure differential is applied. The natural frequencies and mode shapes are shown in Figure 6.8 without any constraint applied to the central bush. Note there is no damping in the bearings: this idealisation was chosen in order to clarify the effect of the central sealing bush.

It is clear from the mode shapes that since the bush is placed at the mid-span of the shaft, it will have essentially no effect on the 'rocking' modes 2 and 4. Owing to the finite axial span of the bush, the influence is not exactly zero, but can be neglected unless pressure differential is very high. Table 6.1 shows the natural frequencies and damping coefficients of modes 1 and 3 for a range of differential pressures across the bush.

Nat freq = 26.7329 Hz   Nat freq = 51.1693 Hz

Nat freq = 173.5959 Hz   Nat freq = 429.9714 Hz

FIGURE 6.8
Natural frequencies and mode shapes.

TABLE 6.1

Effects of Bush on Natural Frequencies and Damping

Pressure Differential (MPa)	Mode 1		Mode 3	
	Frequency (Hz)	Damping Coefficient	Frequency (Hz)	Damping Coefficient
0.2	27.28	0.10	173.44	0.016
0.4	27.83	0.14	173.44	0.023
0.6	28.83	0.17	173.44	0.028
0.8	28.92	0.19	173.44	0.033
1	29.45	0.21	173.44	0.037
2	31.96	0.26	173.45	0.051
5	38.61	0.33	173.47	0.084
10	51.16	>1	173.50	0.12

For mode 1, the 'bounce' mode, as pressure increases to 5 MPa, the natural frequency increases from 26.7 to 38.6 Hz. Note that between 5 and 10 MPa, the nature of the mode changes completely, and it is difficult to make general observations. An important observation, however, is the steady increase in damping as the differential pressure rises. In real plant, the first mode of pumps is often not seen in plant data owing to high damping of this mode. This means that in many cases, the lowest frequency mode observed will be the 'rocking' mode, readily recognisable by the phase difference between the displacements at the bearings. In the simple test sample, the frequency of mode 3 did not change, although the damping shows a steady increase. In real machines, however, the situation is considerably more complex, because there are usually a number of neck-rings and seals, each subject to pressure differentials varying with pump operation.

The influence of the bush on the forced response can be observed from Figure 6.9. An equal imbalance was applied to each of the discs so that response will be governed by the first mode. It is clear that the peak becomes heavily damped.

As illustrated in Table 6.1, the presence of the seals within a pump may have considerable influence on the overall vibrational behaviour of the unit, and furthermore, this effect may vary with time owing to wear within the various clearances. Whilst the shift in critical speeds may be modest, the crucial influence in the behaviour arises from the damping. As wear progresses, damping changes can substantially alter the machine's signature. This wear pattern will also be reflected in the performance characteristics of the pump, that is plots of developed head against flow rate. As so often, the real value of vibration signal analysis is maximised by using the measured data in conjunction with other plant information in forming an overall judgement of a machine's condition.

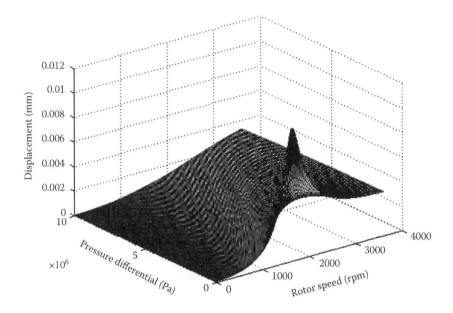

FIGURE 6.9
Response to imbalance.

6.3.2 Other Forms of Excitation

The arguments set out so far outline the influence of the pumped medium on the mass, damping and stiffness properties of the rotor system, but of course, this medium also has a major influence on the dynamic forces arising during operation. The number of vanes on each impeller and the number of channels in the diffuser normally differ by one, usually six and seven, respectively. In a perfect (idealised) pump, each time an impeller blade passes a diffuser vane, there will be some pulse of pressure, resulting in an excitation at 42 times per revolution and harmonics thereof. If the acceleration (as opposed to displacement) is monitored, then a signal is usually present at this fairly high frequency. However, in a real machine, there will normally be sub-harmonics of this ideal frequency. For instance, some erosion or other fault on a diffuser vane would give rise to pulsations at six times per revolution, whereas an imperfection on an impeller blade will yield a frequency corresponding to seven times the shaft speed.

There is a wider context to this discussion: The detailed discussion of centrifugal pumps illustrates the main (and rather obvious) point that the dynamic behaviour of any piece of machinery is determined by the physics of its internal state and operation. Indeed, it is this fundamental connection which gives value to monitoring by means of vibration and performance parameters: the better the understanding of the basic physics of operation, the greater will be the information gleaned from monitoring and diagnostic procedures. The example of large centrifugal pumps has been used as an illustration, since the combination of a high energy density and a fairly light machine implies that the pumped medium has a significant influence on the vibrational behaviour.

6.3.3 Steam Whirl

Many machines show similar effects due to internal fluid motion influences. Steam whirl (a phenomenon related to that observed in oil film bearings) has been observed many times. As reported by Bachschmid et al. (2008), it is often critical, dependent on fine clearances and is extremely difficult to predict with any degree of certainty. This study reports on two identical machines, one of which exhibits steam whirl whilst the other one does not. This contrasting behaviour can only be ascribed to small random in-built differences. Whirl can occur in the glands of some steam turbines, whilst a somewhat different source is the so-called Alford's force, arising from small differences in blade clearances and this is discussed by Adams (2001) and Friswell et al. (2010). It should be emphasised, however, that these excitation sources are really instabilities, rather than true forces. Whilst the analysis of the fluid motion used for the analysis of pump clearances needs some adaption, similar physical concepts may be applied.

6.4 Direct Stator Contact

There are a number of effects of what we may loosely term rotor–stator interaction. The appropriate point at which to begin the discussion is with the rubbing of a rotor against its casing during part of the cycle. This seemingly simple picture is actually very complex and covers a wide range of faults and resultant behaviour. In the first instance, we consider a 'light' rub in which the rotor rubs against its containing stator over a narrow angle of travel during each rotation. This is depicted in Figure 6.10, and we now consider the likely sequence of events. It is assumed that at rest, the rotor and stator are not in contact (apart, i.e. from the bearings), but some whirl due to imbalance causes the rotor to rub against the stator over a limited portion of its orbit.

6.4.1 Extended Contact

6.4.1.1 Physical Effects

The possibilities are well summarised by Muszynska (1989, 2005), but in many instances, there will be insufficient information to fully explain all details, and consequently, it is important to be aware of the various possibilities.

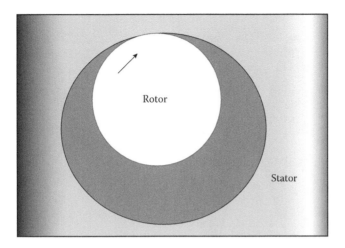

FIGURE 6.10
A rotor–stator rub.

As the rotor contacts the stator

1. Rubbing will cause heat generation, leading to a thermal bend.
2. The change in support will alter the natural frequencies and dynamic response.
3. The impact will generate harmonics (and sub-harmonics).
4. The friction can give rise to torsional excitation.
5. There will be a change in damping.
6. There will be deterioration in plant performance.
7. Acoustic emission (AE) is generated.

6.4.1.2 Newkirk Effect

Since the whirl is synchronous, the same point on the rotor rubs on each revolution, and consequently, heat is generated which causes the rotor to bend.

As discussed in Section 4.3, a bend also gives rise to a synchronous vibration signal, and at constant speed, the effect of this thermal bending is to modify the effective imbalance, and so the vibration will change, in general, in both magnitude and phase. Without embarking on the mathematical details, it is fairly straightforward to see how a cyclic pattern of behaviour develops: because the effective imbalance is changed by the thermally induced bend, the rub conditions and consequently the heating change. This gives rise to a range of possible scenarios which depend on the combination of physical parameters. Clearly, the vibration will change in magnitude and phase over time. This cyclic process was first identified by Newkirk (1926).

Under some conditions, the amplitude will remain within acceptable limits and the result will be a synchronous response whose amplitude varies

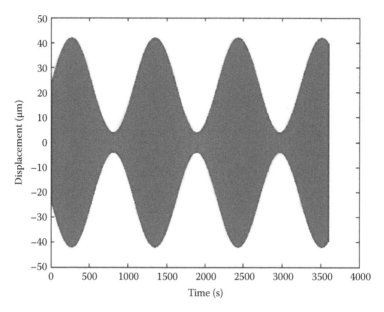

FIGURE 6.11
An example of the Newkirk effect.

slowly over time, and an example of this is shown in Figure 6.11. Various theoretical treatments have been reported (Muszynska 1989) but the difficulty is really the uncertainty of the precise conditions within the machine. Nevertheless, models have been successful in reproducing trends at timescales that correspond well with observations.

Although not shown in Figure 6.11, it is interesting to observe the way in which phase varies with time. Often this reveals a gradual change as the heat developed at the rub causes a thermal bend, which in turn changes the effective imbalance on the rotor. Note that in Chapter 4, it is (strongly) argued that bends and imbalances are not equivalent; nevertheless, taken at a single rotor speed, they can be considered together. The result is, therefore, a gradually changing effective unbalance which may take the form of a change in amplitude, phase or both. From Figure 6.11, it is observed that the timescale associated with this process is about 20 min and long periods like this are typical of large machines. This characteristic time will be determined by a variety of factors, including the dimensions of the machine, thermal conductivity of the rotor and the rotor flexibility. Other factors such as the mass imbalance and the clearance will only influence the extent of the thermal cycling.

There are a range of possibilities covering the interaction of a rotor and its stator:

1. The rotor may rub constantly all around the circumference (although this is a rare condition).

2. The rotor may constantly rub against one part of the stator.

3. The rotor may contact the stator at one portion of its orbit on each cycle: this is the scenario which gives rise to the Newkirk effect as described here.

This seemingly simple process can give rise to a wide variety of consequences, and the parameters of a specific situation will determine which of the terms dominate. For successful diagnosis, the aim is to use a blend of all these factors to yield maximum insight into the problem.

There have been many papers on the topic of thermal rubs in rotating machines, and this is not surprising in view of its importance. Treatments range from the purely empirical to detailed mathematical analyses, which is all a little frustrating in view of the limited available knowledge. It is important to appreciate that there are two quite distinct cases: If there is a short impact between rotor and casing, there will be an exchange of momentum but little heat generation. With a long rub, the situation will be reversed, with little momentum exchange but considerable heat generation. In a real incident, one may observe a combination of these two extremes. Here, we present a description of the rather the slow rub case, but even this yields a complex combination of phenomena.

To develop the discussion, consider the steady-state motion of a rotor which has both imbalance and a bend. The motion is described by

$$\mathbf{K}y + \mathbf{C}\dot{y} + \mathbf{G}\Omega\dot{y} + \mathbf{M}\ddot{y} = \mathbf{K}be^{j\Omega t} + \mathbf{M}\Omega^2\varepsilon e^{j\Omega t} \tag{6.22}$$

The complication with a rub is that the bend is thermally induced and it is time dependent. In all realistic scenarios, however, the thermal timescale is much longer than that of the dynamics, and so it is valid to separate the two problems. As the rotor contacts the stator, heat $\dot{Q}(t)$ will be generated and, although there will be some radiated losses, the net result will give rise to a thermal field within the rotor which will cause a bend. This time-dependent heat generation will give rise to a temperature profile within both rotor and, usually to a lesser extent, the stator. Focussing attention on the rotor, this non-uniform temperature profile gives rise to distortion, tending to bend the rotor. For a given heat input and known losses, the temperature profile of the rotor can, in principle, be calculated using standard methods, either analytical or numerical; but, of course, it is rare to have any precise knowledge of the conditions and consequently, it is far more important to formulate a more general understanding of the processes involved.

The distribution of temperature within the rotor is easily translated into stresses and strains, which in turn result in a combination of forces and moments acting on the rotor, giving rise to the bend term of Equation 6.22. This represents a complicated chain of events which is represented by the chart shown in Figure 6.12.

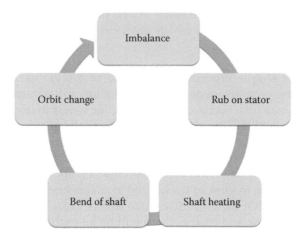

FIGURE 6.12
The Newkirk cycle.

This process does, however, avoid one important question: Does the rotor stay in contact with the stator long enough to generate significant heat, or does it simply undergo an impact and rebound? In either event, there will be significant changes to the dynamic behaviour, but the two scenarios are markedly different.

In reality, any contact will give rise to both types of interactions, with one dominating the other depending on the local parameters. It is worth briefly considering the physics of the interaction: as the rotor makes contact with the stator, two interaction forces arise, one perpendicular to the surface and the other, a friction force, along the surface. Both these forces will be non-linear functions of the relative displacement. The two components will remain in contact for some time, say τ, which will be determined by the local stiffness properties of both components together with the velocity of the rotor.

6.4.2 Collision and Recoil

6.4.2.1 Physical Effects

All rotating machines involve fine clearances between stationary and moving parts, and these are often filled with fluid, which may be the working fluid, coolant or lubricant. These clearances influence the dynamics of the machine in a number of ways which we can categorise into three main divisions. The classification is as follows:

1. Fluid forces predominate – The clearances may act as subsidiary bearings in the manner discussed for neck-ring seals in pumps. A very detailed discussion of the varied phenomena is given by Muszynska (2005). The overall contribution can be the addition of

mass, stiffness or damping. It can also, under certain circumstances, feed energy into the rotor, giving rise to instability in the form of whirl or whip.

2. Rubbing – Leading to a thermal and dynamic cycle.
3. Impact and recoil.

A substantial number of papers on this topic take a variety of models describing the contact between stator and rotor. These involve structural parameters, surface condition and lubrication, and it is clear that the real physics of the processes involved is indeed complex. In a real machine, it is highly unlikely that one would ever have the information to execute a credible calculation. But this is not to diminish the value of this research: the real value lies not in the numbers but rather in clarifying the wide spectrum of effects that can arise from rotor–stator contact.

Edwards et al. (1999) were the first to discuss quantitatively the generation of torsional excitation from rotor–stator contact, and this is, perhaps, a little surprising. It reflects a wide spread tendency to neglect torsion, and this is understandable in view of the difficulties of accurate measurement of torsional motion. As discussed elsewhere, the effort involved in monitoring torsion cannot be justified in many cases, but it remains important to gain an understanding of the phenomena involved.

As the rotor makes contact with the stator, the dynamics of the system is fundamentally changed. The stiffness and damping parameters alter in addition to the introduction of impulsive forces. If the time in contact is short, then these impulsive forces play an important role, whereas for extended contact, discussed in Section 6.4.2, thermal processes assume greater importance. As an illustration of the complexity which can arise with impulsive contacts, an idealised system is now considered in a single direction.

6.4.2.2 Simulation

Now return to the issue of contact between rotor and stator. In addition to the thermal aspects (discussed in Section 6.4.1) of the contact, there is impulse exchange which may occur sufficiently rapid so that the rebound of the rotor is far more significant than the distortion of the motion caused by heat generation. Clearly, the determining factor in this process is the time over which rotor and stator remain in contact, and this in turn depends on the relative stiffness terms as well as velocities.

Once a rotor comes into contact with its stator, the effective stiffness will change. This is indeed a complicated scenario involving impact, torsional excitation and heat generation. Taking just the simplest case, it is instructive to model this as a mass on a spring which is impacting on a second spring as shown in Figure 6.13. At rest, the clearance between the mass and the second spring is δ.

FIGURE 6.13
Idealised one-dimensional impacting.

Free vibration is considered first; for small amplitudes, the displacement is given by

$$x_1(t) = x_0 \sin \omega t \tag{6.23}$$

where
$$x_0 \le \delta$$
$$\omega = \sqrt{k_1/m}$$

Now consider the case when $x_0 > \delta$. Then contact will occur at time

$$\tau = \omega^{-1} \sin^{-1}\left(\frac{\delta}{x_0}\right) \tag{6.24}$$

After the impact has occurred, the displacement is given by

$$x_2(t) = A \cos \omega_2(t - \tau) + B \sin \omega_2(t - \tau) \tag{6.25}$$

where $\omega_2 = \sqrt{(k_1 + k_2)/m}$. By equating position and velocity at the instant of contact

$$A = \delta$$
$$\omega_2 B = x_0 \omega \cos \omega \tau \tag{6.26}$$

Hence

$$B = \frac{\omega}{\omega_2} \sqrt{x_0^2 - \delta^2} = \sqrt{\frac{k_1\left(x_0^2 - \delta^2\right)}{k_1 + k_2}} \tag{6.27}$$

This study can now be continued to establish the 'effective frequency' of this non-linear cycle by determining the time at which the oscillation returns to the zero point. Note that Equation 6.25 can be expressed as

$$x_2(t) = \sqrt{A^2 + B^2} \cos\left[\omega_2(t - \tau) - \tan^{-1}\left(\frac{B}{A}\right)\right] \tag{6.28}$$

Then the maximum deflection occurs at time

$$t_{max} = \tau + \frac{1}{\omega_2}\tan^{-1}\left(\frac{B}{A}\right) \tag{6.29}$$

This is illustrated in Figure 6.14. In this slightly artificial example, the base frequency is 1 Hz, whilst the stiffness $k_2 = 4 \times k_1$; the vibration has an initial velocity corresponding to unit amplitude, and contact occurs at an amplitude of 0.8.

Hence, zero displacement will occur after time $2 \times t_{max}$, and the cycle will be complete at time given by

$$T = \frac{2}{\omega_2}\tan^{-1}\left(\frac{B}{A}\right) + 2\tau + \frac{\pi}{\omega_1} \tag{6.30}$$

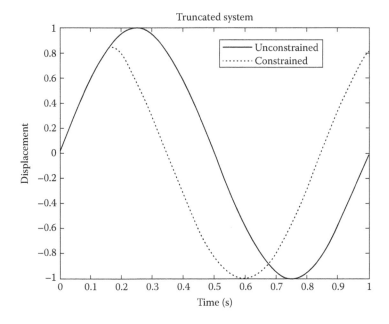

FIGURE 6.14
System with an impact.

The effective frequency is $\Omega = 2\pi/T$, although this terminology must be used with some care: Ω does indeed specify the fundamental frequency, but this will be accompanied by harmonics, since the displacement function is no longer sinusoidal. This simple model is reflected in the machine with rotor–stator contact which generates harmonics of the shaft rotation frequency. More generally, this is the case with all forms of non-linear behaviour.

This study of the free vibration gives some insight into the mechanics, but the analysis of the forced response is a little more complicated. This is because transients arise each time the support conditions change, and whilst an analysis is in principle tractable, the algebra rapidly becomes too complicated to yield a meaningful physical picture. It is, therefore, more convenient to resort to a numerical simulation. Figure 6.15 compares the 'unconstrained' free vibration with that the case including impacting. It is clear that the cycles are shorter, but the motion is no longer sinusoidal but contains a number of harmonics arising from the impacts.

Referring back to Figure 6.13, the forced equation of motion is given by

$$K(x)x + C\dot{x} + M\ddot{x} = F \sin \omega t \qquad (6.31)$$

In this case, damping has been included, since there is no great simplification in ignoring it. Note however that in this case, a sinusoidal form

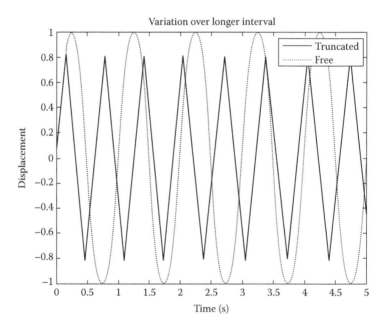

FIGURE 6.15
Free and truncated motion.

for x has not been assumed, since non-linearities are present, giving rise to harmonics.

$$K(x) = k_1 \qquad \text{for } x < \delta$$

$$= k_1 + k_2 \quad \text{for } x > \delta \tag{6.32}$$

If the sample is subject to forcing, a plot of velocity against displacement reveals the harmonic content, and the gradual approach to chaos as δ is reduced. Figure 6.16 shows plots of the response of this example to a unit force at 1.5 Hz with gap values of 80%, 60%, 40% and 20% of the excited amplitude. This provides a very graphic view of the extent of the non-linear behaviour.

A detailed study of the situation by Edwards et al. (1999) makes it clear that for some combinations of speed and geometric parameters, the motion can become chaotic. Muszynska (1989) gives an extensive discussion of the analysis which is indeed rather complex. It confirms that for certain speed and imbalance combinations, chaotic motion can result. Since in the present volume the primary emphasis is on the recognition of fault conditions, attention is restricted to the basic phenomena.

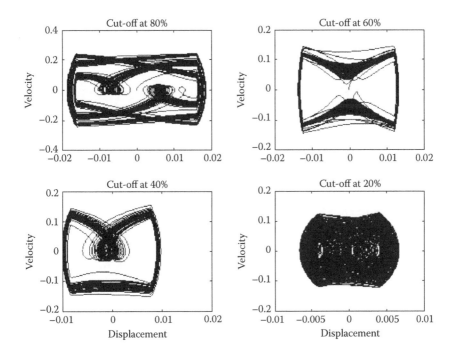

FIGURE 6.16
Phase plot with impacting.

At the most basic level, we consider the motion of a rotor with a single degree of freedom. Note that in the case $x_{max} > \delta$, the simple sinusoidal solution cannot be applied, since the stiffness varies over a cycle. At first sight, one may consider that the two regimes could separately be solved in the frequency domain, but if this route were taken, due allowance would be needed of the transients arising as the mass passes between the two regimes. It is, therefore, most appropriate to consider this problem numerically, and this can be achieved using state-space notation. Equation 6.32 is re-written in the form

$$\begin{bmatrix} 0 & K \\ K & C \end{bmatrix} \begin{Bmatrix} x \\ \dot{x} \end{Bmatrix} + \begin{bmatrix} -K & 0 \\ 0 & M \end{bmatrix} \frac{d}{dt} \begin{Bmatrix} x \\ \dot{x} \end{Bmatrix} = \begin{Bmatrix} 0 \\ A \sin \omega t \end{Bmatrix} \tag{6.33}$$

This may be conveniently represented in non-dimensional units as

$$\begin{bmatrix} 0 & \omega_n^2 \\ \omega_n^2 & 2\xi\omega_n \end{bmatrix} \begin{Bmatrix} x \\ \dot{x} \end{Bmatrix} + \begin{bmatrix} -\omega_n^2 & 0 \\ 0 & 1 \end{bmatrix} \frac{d}{dt} = \begin{Bmatrix} 0 \\ \dfrac{A}{M} \sin \omega t \end{Bmatrix} \tag{6.34}$$

where A has been divided by the mass term, but since at this stage it is an external parameter, no additional symbol is necessary. At this point, some readers may question the need for this step but suffice it to say that it proves to be a convenient notation: some of the reasons for this are outlined in Chapter 3.

These figures show increasing complexity, degenerating into chaos in some cases, and these features are characteristic of rotor–stator contact. Precise theoretical representation of the dynamics is extremely difficult as it is dependent on a number of detailed parameters, many of which will be unknown. Nevertheless, the examples shown earlier do illustrate some general trends which are extremely helpful in the diagnosis of rubbing problems.

6.4.3 Acoustic Emission

When contact does occur, this can be a source of high-frequency oscillations which is somewhat loosely termed 'acoustic emission' by many authors. As rotor and stator impact, there will indeed be a range of high-frequency modes excited and the precise mechanism bringing this about will be dependent on the stiffness profile of the two bodies. Furthermore, the process will be non-linear involving something approaching Hertzian contact (albeit involving cylindrical rather than spherical geometry). In addition to the set of high-frequency modes excited, as the surfaces make contact, surface asperities will distort and break, and it is these processes which give rise to elastic waves in both rotor and stator. The effective forcing function is wide band, and consequently, the observed AE frequencies will reflect the resonances of the various components. Price et al. (2005) were among the first to use the frequency content of AE signals: a more common approach is simply to monitor the number of 'events' as a measure of acoustic activity. Through this route, the monitoring of AE has become a valuable tool in the monitoring of static structures, but it is only in recent years that its applicability to rotating plant has become appreciated.

Figure 6.17 shows a typical signal of the high-frequency vibration. Note these signals are very low in amplitude and need normally 40–60 dB

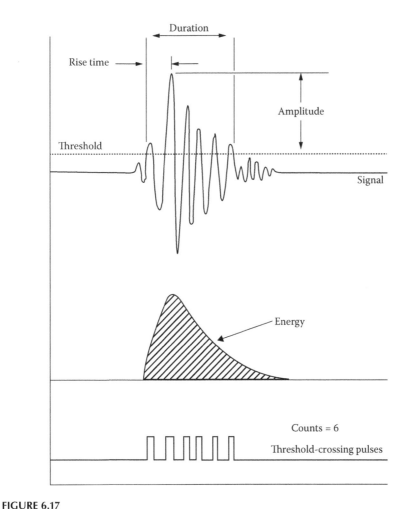

FIGURE 6.17
Typical acoustic emission signal. (Reproduced from Price, E.D. et al., *Proc. Inst. Mech. Eng., Part J – J. Eng. Tribol.*, 219(2), 85, 2005. With permission.)

of amplification. The figure shows a typical transient event, which may be characterised by a number of parameters which are illustrated in the figure. Conventionally, studies in AE have focused on counting the number of so-called events, the number of time the signal exceeds some predetermined threshold value. Whilst this approach has proved very successful in the monitoring of static structure, such as bridges, it is the author's belief that AE technology has considerably more potential for greater use in the future.

True AE signals differ from normal vibration measurements in that they arise as a result of mechanisms within the material, such as the breaking of molecular bonds. They occur at high frequencies, normally in the range of 25 kHz–1 MHz, but these frequencies do not correspond to any on the molecular scale as is sometimes thought (erroneously!). The frequencies corresponding to molecular bonds are at least six or seven orders of magnitude above those of AE, and therefore, it is worth considering the origin of the signals observed. As any event on a molecular scale, such as the breaking of bonds, occurs suddenly on the timescales being considered, the effect on the body is effectively an impulse. This means that a full spectrum of frequencies is generated. Consequently, a stress wave travels through the body, some undergoing reflections and absorption. The components of the wave that are reinforced by the various reflections correspond to the resonant frequencies of the body, or rather, the partitions of the body between reflecting components. Hence, the frequency composition of AE signals reflects the dimensions of the component within which the acoustic activity is occurring. The analysis of the complete AE waveform is a relatively new use of AE, but it is an area which can bring substantial benefits in the study of rotating machinery.

Consider now the processes which may give rise to AE activity. These may be enumerated as follows:

1. Initiation or growth of cracks
2. Plastic deformation
3. Rubbing surfaces (with deformation and breaking of asperities)
4. Turbulent flow in bearings
5. Effects of the working fluid

Several papers have reported the use of AE for the detection of rotor rub, and this provides a sensitive and useful monitor of rotor rubbing. Similarly, studies have been undertaken to detect component cracking in ball bearings (Price et al. 2005).

Any form of rubbing between rotor and stator may be expected to produce both high-frequency vibration and AE, noting here the fundamental difference between the two. The AE is produced as surface asperities distort and break, sending a stress wave through the solid. Leahy et al. (2006) have applied AE techniques to the detection of rubbing phenomena.

6.5 The Morton Effect

A superficially similar effect can occur in rotors with substantial overhung components (such as compressors), yet the phenomenon is rather different in that there are no harmonics or AE generated but simply a cyclic synchronous term. This is known as the Morton effect and was first observed in 1976. Again the root cause of the fluctuating synchronous vibration is the formation of a thermal bend, but in this case, no contact occurs between rotor and stator. Heat is generated not by friction but by the shearing of the oil film at the point of minimum film thickness and, consequently, maximum shear. Because there is no contact, there is no change in dynamic properties and no torsional excitation. Simulations of this effect have been undertaken by Keogh and Morton (1993), Kirk and Guo (2005) and others, but it remains a difficult effect to predict with any accuracy, since it is dependent on a fine balance of conflicting factors.

De Jong (2008) reports work on a large compressor which exhibited the Morton effect. He discusses a range of corrective actions ranging from substantial design changes to simply changing the lubricating oil. Unlike the Newkirk effect, the Morton effect is associated only with oil film bearings, whereas an actual rub can occur on any rotating machine.

6.6 Harmonics of Contact

The essential point about rotor motion within a clearance is that the effective stiffness of the rotor increases sharply as the clearance is taken up; that is the effective stiffness is displacement dependent and the restoring force is a non-linear function of the displacement. An idealised form of this is expressed in Section 5.6.

In reality, nothing is truly linear, but linear analysis helps give considerable insight and is very much simpler than a full non-linear treatment. Following the logic of Section 5.6, we take an equation of the form

$$m\ddot{u} + k_0 u - k_1 u^3 = Fe^{i\omega t} \tag{6.35}$$

Letting

$$u(t) = \sum_{n=-\infty}^{\infty} a_n e^{in\omega t} \tag{6.36}$$

Then substituting this expression into Equation 6.35 gives

$$\sum_{n=-\infty}^{\infty}\left(-a_n mn^2\omega^2 + a_n k_0\right)e^{in\omega t} - k_1 \sum_{n=-\infty}^{\infty} a_n^3 e^{3in\omega t} = Fe^{i\omega t} \tag{6.37}$$

Whilst this is moderately complicated, it is clear that the coefficients a_n are inter-related. For all values of $n \neq 1$

$$\left(-a_{3n}m9n^2\omega^2 + a_{3n}k_0\right) - k_1 a_n^3 = 0 \tag{6.38}$$

Hence, various harmonics of the forcing frequency arise, but the precise details of the amplitudes are not of concern here. However, the presence of the harmonics is of great interest (and often concern) in the monitoring of rotating machinery. In the case presented, imbalance will give rise to a synchronous term plus a series of harmonics as given by Equation 6.37.

This can often be detected graphically in rotor orbits: consider the plots shown in Figure 6.18. The first figure shows a simple orbit which in this case

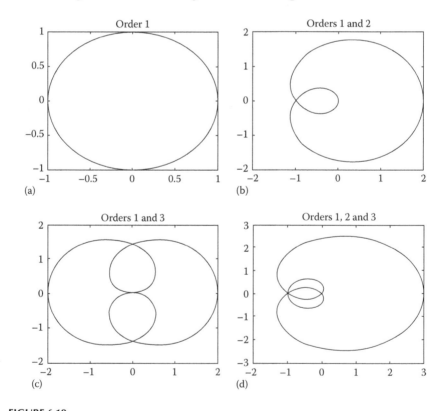

FIGURE 6.18
Effects of harmonics. (a) System with only single frequency, (b) system with base and first harmonic, (c) system with base and second harmonic and (d) system with three frequency terms.

is circular. With the presence of a second harmonic, as depicted in Figure 6.18b, a loop appears, the size of which reflects the amplitude of the second harmonic term. Figure 6.18c and d show that further loops arise with higher harmonics. This extremely simple view of the orbit provides a sensitive indicator of non-linearity: the non-linear terms give rise to harmonic excitation, which in turn generate a looped structure in the rotor orbits.

6.7 Concluding Remarks

The interactions between rotor and stator are complex and often involve details of the machine's operation. In this chapter, the phenomena have been categorised as outlined in the introduction, and in many respects, the behaviour observed will be entirely satisfactory. It is in the small minority of cases where deviations from normal behaviour give indications as to the internal operation and condition of the machine. It is in such cases where an understanding of the internal operation is invaluable to fault diagnosis. Although several analyses are offered here, almost always, there will be little, if any, precise data on which meaningful calculations can be based. The real value of the modelling studies is to gain an appreciation of sensitivities of factors influencing vibrational response. The real aim is the recognition of fault symptoms in order to initiate a more rigorous investigation.

Problems

6.1 The static load on each of the two bearings of a machine is 50 kN. The axial length of each bearing is 0.25 m, the journal diameter is 0.5 m and the diametral clearance is 0.7 mm. Determine the eccentricity using oil of 0.04 Pa s. Calculate the viscosity required to give an eccentricity 80% of the radius.

6.2 The bearings in Problem 6.1 support a uniform steel shaft with a 3 m span. The shaft has two balance planes 1 m inboard of each bearing with 10 g at 0.3 m radius on plane 1 and 5 g at 0.3 m on plane 2. Calculate the response. How does this change if the bearing clearances double with wear?

6.3 A four centrifugal pump develops a head of 200 bar at 5000 rpm. To a first approximation, it may be considered as a uniform shaft 0.2 m diameter, 3 m between rigid short bearings. Each impeller has a mass of 50 kg. If the neck-rings between each stage are 50 mm long with a radial clearance of 0.2 mm, determine the real and imaginary parts of the eigenvalues at 5000 rpm.

6.4 With different valve settings, the pressure developed by the pump in Problem 6.3 is 100 bar. How does the response change?

6.5 A machine shaft passes through a clearance of 0.5 mm. The calculated natural frequency is 38 Hz, but it is observed to be 40 Hz. Estimate the response within the clearance.

References

Adams, M.L., 2001, *Rotating Machinery Vibration: From Analysis to Troubleshooting*, Marcel Dekker, New York.

Bachschmid, N., Pennacchi, P. and Vania A, 2008, Steam-whirl analysis in a high pressure cylinder of a turbo generator, *Mechanical Systems and Signal Processing*, 22, 121–132.

Black, H.F., 1969, Effects of hydraulic forces in annular pressure seals on the vibration of centrifugal pump rotors, *Journal of Mechanical Engineering Science*, 11, 206–213.

Childs, D.W., 1993, *Turbomachinery Rotordynamics: Phenomena, Modeling and Analysis*, Wiley, New York.

De Jong, F., 2008, The synchronous rotor instability phenomena – Morton effect, in *Proceedings of the 37th Turbomachinery Symposium*.

Edwards, S., Lees, A.W. and Friswell, M.I., 1999, The influence of torsion on rotor-stator contact in rotating machinery, *Journal of Sound and Vibration*, 137, 463–481.

Friswell, M.I., Penny, J.E.T., Garvey, S.D. and Lees, A.W., 2010, *Dynamics of Rotating Machines*, Cambridge University Press, New York.

Hamrock, B.J., Schmid, S.R. and Jacobson, B.O., 2004, *Fundamentals of Fluid Film Lubrication*, Marcel Decker, New York.

Keogh, P.S. and Morton, P.G., 1993, Journal bearing differential heating evaluation with influence on rotordynamic behaviour, *Proceedings of the Royal Society: Mathematical and Physical Sciences*, 445, 527–548.

Kirk, R.G. and Guo, Z., 2005, Morton effect analysis: Theory, program and case study, in *Third International Symposium on Stability Control in Rotating Machinery*, Cleveland, OH.

Kramer, E., 1993, *Dynamics of Rotors and Foundations*, Springer-Velag, Berlin, Germany.

Leahy, M., Mba, D., Cooper, P., Montgomery, A. and Owen, D., 2006, Experimental investigation into the capabilities of acoustic emission for the detection of shaft-to-seal rubbing in large power generation turbines: A case study, *Proceedings of the Institution of Mechanical Engineers Part J – Journal of Engineering Tribology*, 220, 607–615.

Muszynska, A., 1989, Rotor to stationary element rub-related vibration phenomena in rotating machinery: Literature survey, *Shock and Vibration Digest*, 21, 3–11.

Muszynska, A., 2005, *Rotordynamics*, CRC Press/Taylor & Francis, Boca Raton, FL.

Newkirk, B.L., 1926, Shaft rubbing: Relative freedom of rotor shafts from sensitive-ness to rubbing contact when running above their critical speed, *Mechanical Engineering*, 48, 830–832.

Price, E.D., Lees, A.W. and Friswell, M.I., 2005, Detection of severe sliding and pitting fatigue wear regimes through the use of broadband acoustic emission, *Proceedings of the Institution of Mechanical Engineers Part J – Journal of Engineering Tribology*, 219, 85–98.

Schweitzer, G., 2005, Safety and reliability aspects for active magnetic bearing applications – A survey, *Proceedings of the Institution of Mechanical Engineers Part I – Journal of Systems and Control Engineering*, 219(16), 383–392.

7
Machine Identification

7.1 Introduction

This chapter is a little different from earlier chapters in that it reports the use of novel techniques which have appeared over recent years. Whilst several of them have been well validated in both laboratory and operational plants, their complexity has meant an inevitable time-lag in becoming widely used by the community rather than being restricted to researchers in the field. Nevertheless, these ideas indicate the direction in which machine diagnostics will progress in the foreseeable future. The bases of the key procedures are outlined and the reader is guided to the references for further details.

7.2 Current State of Modelling

In the previous three chapters, we have discussed methods of analysing vibration signals and surveyed the signature types which are generated by a range of basic faults. We have also discussed various techniques of machine modelling and in situations where the models are sufficiently accurate, it is clear that such models may prove very useful in mapping symptoms to causes, hence providing a useful diagnostic tool. Even prior to the advent of digital computers, some simple models provided significant insight into the dynamics of machinery. This trend can be traced back to the work of Jeffcott (1919) and Stodola (1927). In the last 30 years, more sophisticated numerical models have been developed. The earlier forms were based on the transfer-matrix concept, but in recent years, the finite element (FE) approach has become predominant. The transfer-matrix method, whilst leading to a wildly non-linear computation of critical speeds, had the virtue of making very efficient use of computer storage.

Most modern approaches are based on the FE method, an approach applied to rotor by several authors in the 1970s (see, e.g., Nelson and McVaugh 1976, Nelson 1980). Since the introduction of these techniques, increasingly sophisticated models have been applied to understand observed plant behaviour, and a number of highly beneficial applications have been reported. There are, however, a number of limitations in current models, and some of the reasons for this are discussed in this chapter.

At this point, some general points on modelling are appropriate. As a generic term, modelling may cover a wide variety of tools ranging from a statistical digest of the operational behaviour of a machine to a physics-based mathematical description. In essence, it is a construct to aid understanding of plant operation. The FE method has been applied to rotor systems for many years to gain understanding, and the location of some faults and a basic formulation are given in Chapter 3 and discussed more fully by Friswell et al. (2010). Whilst the statistical approach will be expected to cover the range of operation with some degree of accuracy, it is limited in terms of predictive capability: expressed in more mathematical terms, it can interpolate but not extrapolate, and indeed, this is the case with all empirical models, including artificial neural networks.

Physics-based (almost invariably FE) models do, in principle, have predictive capability. The important question is their fidelity to the real plant and the extent to which fine details of construction are adequately reflected in the models. In many cases, the required details are not known. This variability can be seen in instance where nominally identical machines operate; because they are built to the same drawings, the models for these machines are identical. However, it is almost invariably the case that although the machines show strong similarities in their behaviour, there are subtle differences. In Section 7.4, the sources of these variations are discussed, and in the remaining part of the chapter, some strategies for overcoming these difficulties are outlined.

It should be stressed at this point that this is an important practical issue. All machine operators have a vast amount of data relating to steady on-load running. As described in Chapter 2, these data usually reveal variation with time owing to minor changes in various parameters, but very often, these variations are of the same order of magnitude (or lower) than the uncertainty in modelling, and consequently, little meaningful information can be gleaned from this significant resource of data.

A number of difficulties have been outlined which explain the lack of agreement between models derived purely from engineering drawing and the observed behaviour of actual plant. This leads to the conclusion that, given the benefits of accurate models, the monitored plant behaviour must play some part in the development of these models. There are, however, a number of ways in which this problem may be addressed.

One approach would be to develop a purely empirical model to relate applied forces to response. For a machine, this may take the form of the response to a series of imbalance distributions, but such an approach would involve considerable expense. This could be made somewhat more practical by utilizing the approach described in Section 7.5.1 to calculate the forces, and then calculate the parameters to give optimum fit between model and observations. This fitting may then be obtained by an autoregressive-moving-average (ARMA) approach (see, e.g., Worden and Tomlinson 2000).

Used correctly, this approach can yield effective models, but it is difficult to relate individual terms to physical parameters. This can be overcome by using a physics-based approach in which a model is developed in which specific parameters are adjusted to give best fit – although, as discussed in the next two sections, this is open to differing interpretations.

Section 7.5 gives, in some detail, the basic method proposed by the author and his colleagues, with subsequent sections giving a number of refinements. In this method, a good model for the rotor is assumed, whilst the stiffness, mass and damping matrices representing the foundation are to be determined. This means that each element of these three matrices is a parameter to be determined by the least-square (LS) process. As described, the LS optimization is applied in a single step. Other authors have applied Kalman filtering to this problem, as discussed in Section 7.5.3. The use of the filtering approach has many attractions but also some disadvantages and it is discussed by Provasi et al. (2000).

7.3 Primary Components

To address the issue of model uncertainty and possible tuning models to specific plant items, consider the factors involved in the representation of a machine. It is convenient for our discussion to divide the system into three main components, although it is perhaps more accurate to term these as groups of components. These three are

1. The rotating element
2. The bearings and other interactions between rotor and stator
3. The stator and its supporting structure

Let us now consider each of these in turn. The complete rotating element will, in the case of most turbo-machinery, be a (fairly complex) beam-like structure. This may be represented in terms of an FE model, by some type of beam elements and discs. Appropriate gyroscopic terms are readily included. Experience has developed over the years to the point where rotors may be represented with a reasonable degree of accuracy. There are a number of points to take into consideration, such as the effects of shrink-fit stresses, but the experienced analyst can produce good rotor models. Furthermore, these models may be readily compared with numerical model, by suspending the rotor in a sling and measuring the response of the rotor to either a shaker or impact test. This technique has been reported by Mayes and Davies (1976) and Lees and Friswell (2001), among others. The response measured in a sling gives a very good approximation to the free-free modes of the rotor, thus providing a point of reference to the rotor model.

Unfortunately, this type of independent check is not readily available in the case of the other two components of our global model. In considering the interaction with the stator, the overall phenomena will, of course, be highly dependent on the type of machine in question. The one ever-present contribution will be that of the bearings, and indeed, these will normally represent the dominant force in locating the rotor. Seals and neck-rings also have an important contribution and these have been outlined in Chapter 6. The book by Childs (1993) has a much more extensive discussion on these effects. In discussing large, heavy machinery, the bearings will usually be oil journal bearings, although there are now some large compressors being mounted on active magnetic bearings. These journal bearings are non-linear in their properties and there is some uncertainty as to their behaviour. However, significant strides have been made in recent years and the performance of journal bearings for small oscillation about the equilibrium position is adequately understood. The one limitation here is that the bearing stiffness and damping coefficients depend strongly on the static load on the bearing. With a simple two bearing machine, this presents no problems, but in a complex, multi-bearing situation, there is currently no generally accepted means of knowing the load on each bearing.

Whilst this limits the knowledge of the dynamics, it is by no means the major source of uncertainty. Unless the machine in question is misaligned to a dramatic extent, it is known that the bearing will give a positive direct stiffness at a given speed, and the magnitude of this together with cross-terms can be evaluated albeit approximately. However, the force transmitted via the bearings is dependent on the supporting structure and in this instance, the analyst has little or no idea what the dynamic properties are. These supporting structures often have their own dynamic characteristics and may show resonances below shaft running speed; therefore, it is unknown whether the structure will give a positive or negative restoring force at any given speed. For turbo-generators, this problem has come to the fore in the last 40 years as machines have dramatically increased in size. Foundation structures were at one time high tuned, that is they had no resonances below running speed. As the machines became bigger, it became increasingly difficult to retain adequate rigidity and there was a move towards low-tuned, steel foundations for large machines and these offered a number of practical and financial attractions. It should also be noted that large concrete foundations may also be low tuned. Provided adequate precautions are taken in the avoidance of resonances (or, more precisely, critical speeds) close to operating speeds, such machines are very successful, but there is added difficulty in producing an adequate machine model and in interpreting the machine vibration behaviour. In fact, the representation of the support structure is now believed to be the most significant defect in the modelling of large turbo-machinery.

For a time, it was thought that this problem could be overcome by modelling the support structure using FE techniques. In the period 1970–1990, several groups around the world attempted this, but the success was somewhat limited. The extent of agreement is fairly poor. Whilst it may be argued that the models have predicted the overall trends correctly, it is clear that the agreement is nowhere near good enough for the development of a usable model.

The use of computer methods for the study of machine foundations was reviewed by Lees and Simpson (1983). Their paper concluded that although FE studies could be helpful in the design process, by identifying global trends, far greater precision would be required if a model were to be used as a part of monitoring/fault diagnosis. A few years later, Pons (1986) reported an extensive study on concrete foundations in which extensive measurements were made during power station construction. Once again poor correlation was reported between measured FRF and model predictions. Furthermore, the acquisition of plant data in this way is extremely costly, not only in terms of the effort involved but also more significantly, in terms of the loss of production time on the plant. The situation is exacerbated by the subtle but significant differences between nominally identical units.

7.4 Sources of Error/Uncertainty

Whilst in principle, the effects due to the supporting structure are easily included in a model, in practice, it is impossible to obtain adequate data to formulate the model. The supporting structures often include bolted and welded joints, sliding keyways and expansion joints, multi-various pipework and variety of ancillary equipment. In addition to these factors, there are minor build differences which lead to marked changes in vibrational behaviour. A simple example of this may be illustrated by considering the modal behaviour of a portal frame such as that shown in Figure 7.1. Portal frames such as this form part of many machine support structures, but there are a number of decisions to be made as to the modelling of it. Some of these are as follows:

1. Can a beam representation be used or is solid modelling required? In the latter case, the effort required to generate the model is significantly increased.

2. If beams are used, where is the neutral axis (the usual assumptions of centre lines may distort rotational effects at the joints).

3. How are the joints treated?

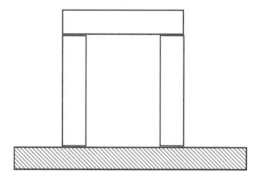

FIGURE 7.1
Portal frame structure.

Hirdaris and Lees (2005) discuss the problems of modelling a portal frame, and Oldfield et al. (2005) give a more general discussion of the modelling of joints.

As with any large, complex structure, there are numerous difficulties in modelling the structure. The conventional modelling units, beams and plates are in themselves idealisations of practical geometries. Yet, fully three-dimensional elements present a prohibitive problem of data preparation and computing. However, the real problem goes much deeper than this; in virtually all practical cases, there is insufficient data to specify the structural properties. The other problems arising for the modeller may be listed as follows:

- Appropriate representation of individual components
- The influence of machine attachment points including keyways
- The contribution to stiffness of machine casings
- The influences of ancillary machinery attached to the support structure

By reference to Figure 7.1, some of the difficulties of modelling become readily apparent. For a slender structure, the frame may be adequately represented by three beams. The corner joints are represented as constraints to a 90° angle with the vertical and horizontal members. With such a model, the natural frequencies and mode shapes are readily calculated.

As the thickness of the beams becomes greater in proportion to their length, the modelling option becomes much less clear. Whilst it is probably acceptable to make representation in terms of beams on the centre line, it is clear that problems arise over the modelling of the corners. Firstly, the mass distribution is inappropriate and secondly, perhaps more important, the corners have now some shear characteristics, and the joint can no longer be represented as simply two beams meeting at right angle.

The portal frame shown in Figure 7.1 is typical of many which form part of the supporting structure of turbo-alternators. In this instance, we consider the top to be rigid and of significant mass, but the flexural rigidity of the two vertical columns differs by 1%. Such a discrepancy is encountered in real plant; indeed, it would be difficult to ensure consistency to within much smaller tolerance. To put this in context, assuming that the columns are a rectangular box section, the wall thickness must be consistent to 1% or the box width to 0.33% – a demanding specification.

7.5 Model Improvement

There are very real practical difficulties in obtaining the required data on turbine foundations. A very convenient way of measuring the dynamics of many structures is through the use of a calibrated hammer which applies an impulse to the structure under study. Provided that the impact is sharp enough, this form of excitation provides a wide spectral range. Whilst there are a number of practical problems associated with the technique, it is generally convenient for site use and usually more convenient (although somewhat less accurate) than, for example, electro-magnetic shakers, the problems are associated with frequency resolution and signal/noise ratio.

In the case of large turbo-set foundations, however, the usefulness of the impact approach was somewhat limited. Whilst there has been little detailed study of this issue, Lees and Simpson (1983) reported that the difficulty lies in obtaining adequate signal-to-noise ratios to enable the determination of cross-receptances, particularly in the lower part of the frequency range. This, added to the large modal density, dictates that any detailed study would have to be performed using a shaker. However, this implies further problems. Clearly, a large exciter, applying sinusoidal forces of order 10 kN, is required for large power turbines, and this implies a test which is laborious to plan and execute. This creates a further difficulty in the time requirements: It is precisely because the plants have a very high earning capacity that accurate models are required for the rapid diagnosis of faults. But this also means that time on the plant is at a premium and it will rarely be cost beneficial to carry out extensive vibration surveys.

A further difficulty arises because the plant will never be in the ideal state for testing. To examine the dynamics properties of the supporting structure, the requirement is for the machine to be complete with all its casings but no rotors in place. This is important, since the upper casings can contribute significantly to the overall stiffness of the assembly. This is a configuration which will seldom/never arise during normal operation, and furthermore, it would prove extremely expensive to provide these conditions. To put these aspects into correct perspective, it is important to observe that large efficient machines may have a replacement cost of order £300,000 per day or more; their high value on the one hand dictates the importance of having accurate models but also severely limits the availability for plant testing.

Kramer (1993) has suggested that the influence of foundations may be adequately modelled within an overall system, but the difficulty lies in the choice of appropriate parameters. With relatively stiff concrete foundations, reasonable answers may be obtained, but for low-tuned steel structures, the problems in obtaining an accurate representation are very much greater.

The high cost of plant implies that the only real option to obtain the required data is to access the data whilst the machine is in operation, but this poses a number of questions as to the appropriate way forward. One might consider the application of an external force to the machine during operation, but few plant managers would sanction such a procedure. The remaining possibility is the use of plant data directly. There are two closely related approaches to this which will now be discussed.

7.5.1 System Identification

At this point, it is important to distinguish between direct and inverse problems. To use a relevant example, consider the equation of motion of a system under the action of a sinusoidal force or set of forces. This can be expressed by the usual equation

$$\mathbf{K}x + j\omega\mathbf{C}x - \mathbf{M}\omega^2 x = \mathbf{F}_0 e^{j\omega t} \tag{7.1}$$

Clearly, if the three matrices are known, then for any set of forces $\mathbf{F}_0$, the resulting displacement vector is readily calculated. This is a direct problem and is reasonably straightforward. This is so because usually it is clear from the physics of the situation that there will be a clear and unique answer.

The inverse problem to that just posed would be the determination of the matrices $\mathbf{K}$, $\mathbf{C}$, $\mathbf{M}$, given the measured response to a given set of forces. In contrast to the direct (or forward) problem, it is not clear that there is a unique answer, and in general, the approach to inverse problems is somewhat more involved. It is worth noting that in the direct problem, there are n parameters to be determined (just the number of degrees of freedom), whereas in the inverse problem, the number of unknowns is $3 \times n \times (n + 1)/2$ and so clearly several frequency components must be considered together. But, even given this, the problem is not straightforward. Simply considering a range of frequencies in order to yield the problem, theoretically soluble may not be sufficient in a realistic case in which the signal (vibration) is contaminated with measurement noise. It is, therefore, appropriate to utilize all the information available to yield the most accurate result.

To do this, Equation 7.1 is rewritten in the rather different form

$$\mathbf{WS} = \mathbf{F}_0 \qquad (7.2)$$

In this form, the matrix **W** contains all the measurement and frequency terms, whilst the vector S comprises all the unknown matrix elements in some predetermined order. To clarify this, consider the two degree-of-freedom case. Note there will be a total of nine unknown parameters, namely $\{K_{11}, K_{22}, K_{21}, M_{11}, M_{22}, M_{21}, C_{11}, C_{22}, C_{21}\}$. The ordering can be chosen and the reasons for this particular choice will be outlined in due course. The immediate task is to express the equations of motion in the form of Equation 7.2. For each (discrete) frequency component, we may write

$$
\begin{bmatrix}
x_{01} & 0 & x_{02} & -\omega^2 x_{01} & 0 & -\omega^2 x_{02} & j\omega x_{01} & 0 & j\omega x_{02} \\
0 & x_{02} & x_{01} & 0 & -\omega^2 x_{02} & -\omega^2 x_{01} & 0 & j\omega x_{02} & j\omega x_{01}
\end{bmatrix}
\begin{Bmatrix}
K_{11} \\ K_{22} \\ K_{21} \\ M_{11} \\ M_{22} \\ M_{21} \\ C_{11} \\ C_{22} \\ C_{21}
\end{Bmatrix}
= \begin{Bmatrix} F_{01} \\ F_{02} \end{Bmatrix}
$$

$$(7.3)$$

There is one such equation for each of the N_f frequencies and these can be concatenated to give a single equation of the form (Equation 7.2) in which the matrix **W** has the dimension $2 \times N_f \times 9$.

With a large number of frequencies measured, the problem is over-specified, but since the measurements are contaminated with noise, this is desirable and allows the analyst to obtain a 'best estimate' and to assess the errors involved. In general, **W** will be rectangular and hence, an appropriate way forward is to multiply Equation 7.2 by the transpose of **W** to give

$$\mathbf{W}^\mathbf{T}\mathbf{WS} = \mathbf{W}^\mathbf{T}\mathbf{F}_0 = \mathbf{AS} \qquad (7.4)$$

Since **A** is now a square matrix, we may write

$$\mathbf{S} = (\mathbf{W}^\mathbf{T}\mathbf{W})^{-1}\mathbf{W}^\mathbf{T}\mathbf{F}_0 \qquad (7.5)$$

This is called the Moore–Penrose solution and it is easily shown that this is equivalent to the least squares (LS) solution of the set of equations.

At this stage, the problem is, in principle, solved, but there are some difficulties to be resolved. These problems centre on the conditioning of the pseudo-inverse in Equation 7.5. This is closely associated with the level of noise in the measurements: the dichotomy is in seeking to fit measurements to a model, generally increasing the number of degrees of freedom will improve the agreement between model and measurement; too many degrees of freedom will result in the modelling of the noise. This leads to a physically unrepresentative model which, although reproducing the particular data used, has little or no predictive capability.

7.5.2 Error Criteria

In all identification procedures, it is important to include all available data and because of inevitable measurement and other errors, the data will contain some inconsistencies. The problem becomes one of minimizing some measure of the errors, usually the sum of the squares of the individual errors. But even then, there is a choice between the equation error and the output error. From an analytic standpoint, these two at first sight seem equivalent, but this is not the case as the errors are weighted rather differently.

Consider first the errors in Equation 7.1.

At each frequency considered, the equation error can be defined as

$$\varepsilon_1(\omega) = \mathbf{F}(\omega) - \mathbf{K}\mathbf{x} - j\omega\mathbf{C}\mathbf{x} + \omega^2\mathbf{M}\mathbf{x} \qquad (7.6)$$

This is a vector quantity and a total error parameter can be defined as

$$E_1 = \sum \varepsilon_1^T(\omega_i)\varepsilon_1(\omega_i) \qquad (7.7)$$

where the summation includes all frequencies in the transient being studied. This has the great benefit that the resulting LS calculation is linear in all the unknown elements of the matrices $\mathbf{K}$, $\mathbf{M}$ and $\mathbf{C}$, but unfortunately, it can lead to errors, because it can be sensitive to noisy data. This can be appreciated physically by observing that at some frequencies, there may be high forces acting on the foundation but with little response. As displacement is the quantity measured, this may be subject to substantial noise, which can give rise to bias in the estimated parameters. The way around this is to express the error in terms of the output, which is, of course, the measured quantity. This gives

$$\varepsilon_2(\omega_i) = \frac{\mathbf{F}(\omega_i)}{\mathbf{K} + j\omega_i\mathbf{C} - \omega_i^2\mathbf{M}} \tag{7.8}$$

Then

$$E_2 = \sum \varepsilon_2^T(\omega_i)\varepsilon_2(\omega_i) \tag{7.9}$$

Note that for exact data, the two formulations are equivalent, but they weight noise very differently. Equation 7.7 is called the equation error, whilst Equation 7.9 gives the output error. Output error leads to much better estimates of the unknown parameters because of the more appropriate weighting on interference in the data. The substantial penalty paid for this is the mathematical complication arising from the fact that Equation 7.9 is non-linear in all the unknown parameters. In practice, this means that an iterative approach is required to obtain the solution. One approach is to use the equation error method to give an initial solution, and then to use this as a starting point in an iterative scheme.

7.5.3 Regularisation

The process of improving the quality of the results from an inverse problem is known as 'regularisation', and this consists of a range of methods which are used in combinations to suit the particular problem and the aim is to produce a unique, physically meaningful solution.

There are a number of techniques employed to achieve a good result and in this section, three of the most important approaches are described; these are weighting, truncation of variables and singular value decomposition (SVD).

The methods used in regularisation can be conveniently divided into three:

1. Appropriate scaling of parameter and weighting of measurements to minimize errors
2. The elimination of unimportant variables by using any available knowledge of the underlying physics of the problem
3. Use of a stable numerical procedure, such as the singular value decomposition, which is outlined in Section 7.4.4

The first step in seeking to optimise the estimation procedure is to ensure that the matrix elements in Equation 7.3 are all of a similar order of magnitude. At first sight, this is problematic simply because the scale of stiffness and mass terms are very different, but this can be resolved simply by re-scaling the frequency scale. Each frequency ω is divided by the first natural frequency ω_c and then the unknown parameters of Equation 7.3 are transformed to become

$$v = \left(K_{11} \ K_{22} \ K_{12} \ C_{11}\omega_c \ C_{22}\omega_c \ C_{12}\omega_c \ M_{11}\omega_c^2 \ M_{22}\omega_c^2 \ M_{12}\omega_c^2 \right)^T \qquad (7.10)$$

The important point to note here is that all the nine unknowns are of a similar order of magnitude. A very approximate value for the natural frequency ω_c will suffice for this scaling. Equation 7.3 then is re-written as

$$\begin{bmatrix} x_{01} & 0 & x_{02} & -\left(\dfrac{\omega}{\omega_c}\right)^2 x_{01} & 0 & -\left(\dfrac{\omega}{\omega_c}\right)^2 x_{02} & j\dfrac{\omega}{\omega_c}x_{01} & 0 & j\dfrac{\omega}{\omega_c}x_{02} \\[4mm] 0 & x_{02} & x_{01} & 0 & -\left(\dfrac{\omega}{\omega_c}\right)^2 x_{02} & -\left(\dfrac{\omega}{\omega_c}\right)^2 x_{01} & 0 & j\dfrac{\omega}{\omega_c}x_{02} & j\dfrac{\omega}{\omega_c}x_{01} \end{bmatrix} v = \begin{Bmatrix} F_{01} \\ F_{02} \end{Bmatrix}$$

$$(7.11)$$

Having scaled the parameters in this way, the next step is 'weighting' or simply recognising that in practice, some readings are more reliable than others. Typically, at frequencies close to resonance, high vibration levels will result which far exceed any background noise, and consequently, these readings will be relatively accurate and should be assigned a high weighting. Mathematically, this is achieved by returning to Equation 7.2 and multiplying by a diagonal matrix, Π. The equation becomes

$$\Pi W S = \Pi F_0 \qquad (7.12)$$

Following the discussion, the solution is given by

$$S = \left(W^T \Pi^T \Omega \Pi \right)^{-1} W^T \Pi^T F_0 \qquad (7.13)$$

This can prove to be an effective step in improving the accuracy of the estimation procedure, but all available steps are often required to yield accurate results. A most important step is to ensure that an appropriate number of degrees of freedom are selected. It is obvious that to fit any given curve, increasing the number of degrees of freedom in the 'trial function' will enhance the degree of fit. This, however, is not the crucial issue: the problem is that if the fitting curve is given too much freedom, the result will reflect the noise in the measured. To some degree, choosing the correct combination of parameters is an iterative process.

Care is needed to ensure that the appropriate number of variables is included for identification and this in itself has significant discussion in the literature (e.g., Smart et al. 2000). This is a problem for both physics-based models and the more statistically based methods. The problem is that increasing the number of variables will always give an improved fit to the experimental data, but the derived model will have little or no predictive capability. This is because the parameters are 'used' to model the data noise rather than reflecting the true structure of the machine. An effective way around this difficulty is simply to constrain certain variables to be zero and several instances of this are mentioned in the study on machine foundations. This constraint is simply applied by deleting the appropriate rows and columns of the matrices and vectors. With these considerations, Equation 7.13 can be used to obtain appropriate estimates, but care is needed to minimise the effects of measurement errors. Rather than directly proceeding to use the pseudo-inverse matrix as in Equation 7.13, it is preferable to use the singular value decomposition. This is discussed in many textbooks (see, e.g., Golub and van Loan 2012) and just the basic outline is given in the next section and is discussed further in Chapter 8.

7.5.4 Singular Value Decomposition

Using this technique enables the analyst to solve Equation 7.13 as a series of contributions from the set of modal terms. Errors arise in the treatment of modes with very small singular values. This is because the small singular value means a contribution will be magnified and secondly, combinations of variables with small singular values have little influence on the errors in the set of equations. The way around this is to truncate the set of singular values at some minimum value, setting all subsequent terms to this value. To solve the problem by this method, consider Equation 7.12 and decompose the matrix $[\mathbf{\Pi W}]$ which has dimensions $m \times n$ where $m \gg n$, as

$$\mathbf{\Pi W} = \mathbf{UAV}^T \tag{7.14}$$

where
  $\mathbf{U}$ has dimensions $m \times m$
  $\mathbf{A}$ has $m \times n$
  $\mathbf{V}$ has $n \times n$

Both $\mathbf{U}$ and $\mathbf{V}$ are orthogonal matrices, whilst $\mathbf{A}$ contains the singular values on its diagonal, with zeros in all other elements. Hence, re-writing Equation 7.8

$$\mathbf{UAV}^T\mathbf{S} = \mathbf{W}F_0 \tag{7.15}$$

Then, using the orthogonality properties of $\mathbf{U}$ and $\mathbf{V}$, the solution may be recovered as

$$\mathbf{S} = \mathbf{V}\mathbf{A}^{-1}\mathbf{U}^T\mathbf{W}F_0 \tag{7.16}$$

Although $\mathbf{A}$ is a rectangular matrix, it does have an inverse because of its particular structure:

$$\mathbf{A} = \begin{bmatrix} \lambda_1 0 0 0 0 0 \\ 0\lambda_2 0 0 0 0 \\ \vdots \ddots \vdots \\ 0 0 0\lambda_n 0 0 \end{bmatrix} \Rightarrow \mathbf{A}^{-1} = \begin{bmatrix} \dfrac{1}{\lambda_1} 0 0 0 \\ 0\dfrac{1}{\lambda_2} 0 0 \\ \vdots \ddots \vdots \\ 0 0 0\dfrac{1}{\lambda_n} \\ 0 0 0 0 \\ 0 0 0 0 \end{bmatrix} \tag{7.17}$$

This allows the solution of the inverse problem to be expressed in terms of the singular values and their corresponding eigen-functions.

7.6 Application to Foundations

It has been appreciated for a considerable time that the supporting structures play a significant role in determining the dynamic properties of large machines and this has increased in significance with the trend towards flexible low-tuned supports, that is, structures with resonances below the machine's running speed. From an academic point of view, the obvious approach to this problem would be the performance of a full modal test, but this is not feasible for commercial reasons. A full test on these complex structures would require several weeks, and the high cost of lost production deems this route unviable. Another problem is that, for a meaningful test, the machine would need to have all its casings in place but with no rotor present, a configuration which never occurs. An additional issue is that such an extensive series of tests would give information on a single machine, whereas it is now appreciated that even nominally identical machines, in fact, display subtle but important differences in behaviour. A series of detailed tests were carried out during the construction of Bugey Power Station as reported by Pons (1986), and these results have helped to gain insight into the modelling of these structures.

7.6.1 Formulating the Problem

Recognising the problems in this, attention has inevitably been turned to the use of operational plant data to infer system properties. The problem of applying system identification techniques to a rotating machine has received attention over recent years, and several approaches have been developed by groups around the world. In determining the dynamic properties of structures, modal analysis (see, e.g., Ewins 2000) has become the predominant approach. In this method, a known force is applied to a structure at one or more locations and a highly developed methodology. Some elements of this approach have been applied to the identification of machine foundation particularly by Vania (2000) and the subsequent work by Pennacchi et al. (2006). Their general approach is to tune models using a Kalman filtering approach (basically an iterative LS method) to refine an initial modal representation. A number of difficulties arise, but the substantial progress has been made in recent years.

In using the operating machine as the test-bed, the force levels exciting the structure have to be assessed and there are two main ways of doing this: the first option is to measure the relative motion across each of the bearings and then use a bearing model of some form to formulate the force transmitted to the structure (Vania 2000). The alternative route, suggested by Lees (1988), was to use a validated model of the rotor and then calculate the force arising from a known imbalance, considering the dynamics of the (fully determined) shaft. Whilst this route demands some additional calculation, it does have some advantages. Later work by Lees and Friswell (1997) demonstrated that the imbalance could be regarded as an additional parameter (or set of parameters) to be identified.

The other main choice to be made is the choice of parameters for the model: should the choice be physical parameters or modal ones; again there are strong arguments on both sides. Modal models typically have many few degrees of freedom to deal with, but the equations are inherently non-linear, implying the need for some form of iterative solution. This imposes the necessity of an initial FE model of structure in order to provide a starting point for the full calculation. Because each bearing is implicitly treated separately in the calculation, no means of estimating the overall imbalance has yet been suggested using the modal picture.

7.6.2 Least Squares with Physical Parameters

The use of physical co-ordinates is in some ways rather clumsy, involving a larger number of unknowns but offering a clear physical picture with some advantages. Although it becomes necessary to perform a non-linear estimation to refine the answers, the approach leads to a linear estimation which is then used in the iterative loop. Lees (1988) suggested using plant data by considering the rotor itself to be the source of excitation. In this approach, the forces were calculated by relying on a good rotor model then making the assumption that a bearing model is available. A modified form of this approach was later developed as discussed below. The paper sought values for the stiffness and mass matrices of the support on the assumption that machine damping is dominated by the bearings.

The calculation was originally developed for use on older machines where data on absolute shaft location are not readily available. The distinguishing feature of the method of Lees (1988) and the subsequent papers, culminating in Smart et al. (2000), are the use of a detailed rotor model to infer the magnitude of forces acting. The basic concept of the method is that the rotor motion is produced by the action of two sets of forces, each acting at a discrete set of axial locations:

1. A set of bearing forces, which are unknown, acting at the bearing location.
2. A set of unbalance forces acting at a predetermined set of balance planes. These forces may be considered as known by considering the difference between two balance runs. [Note, however, that later work (Lees and Friswell 1997) has shown that this restriction can be removed.]
3. Because the shaft motion, or rather, the related pedestal motion is measured, the bearing forces can be expressed in terms of the unbalance.

Since the forces at the bearings on the rotor are now determined, the forces acting on the structure are equal and opposite, and their effect is measured as the pedestal vibration. The dimension of the vector $\{F\}$ is the number of bearings (times two if both perpendicular directions are considered). The vector $\{F\}$ represents the force acting on the rotor from the bearing reaction; hence, the force acting on the supporting structure is just $-\{F\}$.

In the case of a foundation with many modes within the running frequency range, the structural matrices should vary. This may be treated by subdivision of the frequency range, resulting in a larger LS problem, but no additional difficulty in principle.

This approach will now be considered in some detail. With the usual assumption of linearity, the equation of motion can be written as

$$\mathbf{K}x + \mathbf{C}\dot{x} + \mathbf{M}\ddot{x} = \mathbf{F} \tag{7.18}$$

Written in this form, the equation is not very helpful inasmuch as there is only incomplete knowledge of the stiffness and mass matrices. Consider the purely sinusoidal excitation: now Equation 7.18 can be expressed as

$$(\mathbf{K} + j\omega\mathbf{C} - \omega^2\mathbf{M})\mathbf{X}_0 e^{j\omega t} = \mathbf{Z}(\omega)\mathbf{X}_0 e^{j\omega t} = \mathbf{F}_0 e^{j\omega t} \tag{7.19}$$

where $\mathbf{Z}$ is the dynamics stiffness, part of which is unknown. However, the parts which are known can be utilized to infer the remaining terms, and to achieve this, after cancelling the sinusoidal terms in Equation 7.19, the dynamic stiffness is partitioned and the equation is re-written in the form

$$\begin{bmatrix} \mathbf{Z}_{R,ii} & \mathbf{Z}_{R,ib} & 0 \\ \mathbf{Z}_{R,bi} & \mathbf{Z}_{R,bb} + \mathbf{Z}_B & \mathbf{Z}_B \\ 0 & \mathbf{Z}_B & \mathbf{Z}_B + \mathbf{Z}_F \end{bmatrix} \begin{Bmatrix} \mathbf{X}_{R,i} \\ \mathbf{X}_{R,b} \\ \mathbf{X}_{F,b} \end{Bmatrix} = \begin{Bmatrix} \mathbf{F}_u \\ 0 \\ 0 \end{Bmatrix} \tag{7.20}$$

The dynamic stiffness terms $\mathbf{Z}_R$, $\mathbf{Z}_B$ and $\mathbf{Z}_F$ refer to the rotor, bearing and foundation, respectively. Experience has shown that the rotor can be modelled with good accuracy and furthermore, the adequacy of any rotor model may be verified by impulse or shaker tests in works; therefore, $\mathbf{Z}_R$ can be regarded as known. The subscripts i, b refer to internal and bearing locations. The bearing may also be modelled; although some difficulties arise, it can be shown that the uncertainties have only a limited influence on the overall result, namely, the estimate of the foundation dynamic stiffness, $\mathbf{Z}_F$.

There are several ways to proceed from Equation 7.18, depending on what plant measurements are available. Here, the case considered relies solely on measurements of bearing pedestal vibration; whilst this is simplest from a measurement perspective, it does involve a little more mathematical manipulation. This is because the terms $\mathbf{X}_{R,ii}$ and $\mathbf{X}_{R,ib}$ have to be eliminated from 7.20. These two parameters can be expressed as

$$\mathbf{X}_{R,b} = -(\mathbf{Z}_{R,bb} + \mathbf{Z}_B)^{-1}(\mathbf{Z}_{R,bi}\mathbf{X}_{Ri} + \mathbf{Z}_B\mathbf{X}_{Fb})$$

$$\mathbf{X}_{R,i} = \mathbf{Z}_{R,ii}^{-1}(F_u - \mathbf{Z}_{R,ib}\mathbf{X}_{R,b}) \tag{7.21}$$

With these substations, the third line of Equation 7.20 can be expressed as

$$-\mathbf{Z}_B(\mathbf{Z}_{R,bb} + \mathbf{Z}_B)^{-1}\left(\mathbf{Z}_{R,bi}\left(\mathbf{Z}_{R,ii}^{-1}(F_u - \mathbf{Z}_{R,ib}\mathbf{X}_{R,b})\right) + \mathbf{Z}_B\mathbf{X}_F\right) + (\mathbf{Z}_B + \mathbf{Z}_F)\mathbf{X}_F = 0 \tag{7.22}$$

However, this equation contains $\mathbf{X}_{Rb}$ which is unknown. But this can be eliminated using the third equation in Equation 7.20 and writing

$$\mathbf{X}_{Rb} = -\mathbf{Z}_B^{-1}(\mathbf{Z}_B + \mathbf{Z}_F)\mathbf{X}_F \tag{7.23}$$

Substituting Equation 7.23 in Equation 7.22 yields an equation in which the only unknown quantities are $\mathbf{Z}_F$ and $\mathbf{F}_u$.

At first sight, this looks fairly formidable, but it contains only two terms which are unknown, $\mathbf{Z}_F$, the foundation dynamic stiffness and $\mathbf{F}_u$, the imbalance. In fact, working with the difference between two imbalance runs, it is possible to regard $\mathbf{F}_u$ as known. Lees and Friswell (1997) show that the imbalance may simply be treated as a set of additional parameters to be identified, and this is briefly discussed in Section 7.7. Later work (Sinha et al. 2002) demonstrates that the bearings may also be treated in similar manner. By defining some new terms, Equation 7.20 can be expressed as

$$\mathbf{PF}_u + \mathbf{Q} + \mathbf{SZ}_F = 0 \tag{7.24}$$

where the matrices $\mathbf{P}$ and $\mathbf{S}$, together with the vector $\mathbf{Q}$, are rather complicated but well-defined functions of $\mathbf{Z}_R$, $\mathbf{Z}_B$ which are both known and the vector $\mathbf{X}_F$ which is measured. This equation is now re-cast in the form of Equation 7.2 for the identification process. The foundation matrix $\mathbf{Z}_B$ is expanded to its constituent stiffness mass and damping terms and the equation becomes

$$\mathbf{P}_\omega \mathbf{F}_{u\omega} + \mathbf{Q}_\omega + \mathbf{S}_\omega \mathbf{v} = 0 \tag{7.25}$$

where, as discussed in Section 7.5.1, the matrix $\mathbf{S}_\omega$ contains the measured vibration readings, as does the vector $\mathbf{Q}_\omega$, and the vector $\mathbf{v}^T = \{k_{11} \quad \cdots \quad k_{1n} \quad m_{11} \quad \cdots \quad m_{1n} \quad c_{11} \quad \cdots \quad c_{1n}\}^T$ represents the parameters to be determined. The subscript ω is a reminder that this equation is formulated for each frequency of the excitation, and hence, all these result in an over-specified problem of the form

$$\mathbf{Sv} = -\mathbf{PF}_u + \mathbf{Q} \tag{7.26}$$

This formulation is suitable for machines on which only measurements of bearing vibrations are available, but, as shown, some modelling of the bearings is needed, although the results are not very sensitive to bearing errors. On more modern machines, however, measurements of the rotor vibration are available by using proximity probes. In this case, a different, and rather simpler, formulation is used and this is discussed in Section 7.6.2.1. Not only is this simpler, but also the availability of rotor data facilitates access to some further insights into the machine's condition.

7.6.2.1 Formulation with Shaft Location Data

For the case in which shaft motion is also monitored, a rather simpler formulation may be derived from Equation 7.19 to give

$$\begin{bmatrix} \mathbf{Z}_{R,ii} & \mathbf{Z}_{R,ib} & 0 \\ \mathbf{Z}_{R,bi} & \mathbf{Z}_{R,bb} & 0 \\ 0 & 0 & \mathbf{Z}_F \end{bmatrix} \begin{Bmatrix} \mathbf{r}_{R,i} \\ \mathbf{r}_{R,b} \\ \mathbf{r}_F \end{Bmatrix} = \begin{Bmatrix} \mathbf{f}_u \\ \mathbf{f}_b \\ -\mathbf{f}_b \end{Bmatrix} \tag{7.27}$$

At this stage, of course, the force exerted on the shaft at the bearings, $\mathbf{f}_b$, is unknown, but it is easily eliminated and recovered from the calculation later. By adding the second and third rows

$$\begin{bmatrix} \mathbf{Z}_{R,ii} & \mathbf{Z}_{R,ib} \\ \mathbf{Z}_{R,bi} & \mathbf{Z}_{R,bb} \end{bmatrix} \begin{Bmatrix} \mathbf{r}_{R,i} \\ \mathbf{r}_{R,b} \end{Bmatrix} + \begin{bmatrix} 0 & 0 \\ 0 & \mathbf{Z}_F \end{bmatrix} \begin{Bmatrix} 0 \\ \mathbf{r}_F \end{Bmatrix} = \begin{Bmatrix} \mathbf{f}_u \\ 0 \end{Bmatrix} \tag{7.28}$$

It is worth considering each of the terms in some detail: all the values of r are known, since these are the measured quantities. In the first matrix, all four terms $\mathbf{Z}_R$ are known, since this is just the model of the rotor which (it is assumed) has been validated separately. $\mathbf{Z}_F$ is just the model of the foundation which is to be determined. The first row of Equation 7.28 is now used to eliminate the internal rotor degrees of freedom using dynamic condensation. The imbalance force may be either known (by taking the difference of two balancing runs) or treated as an additional unknown vector to be determined.

In general, the more measurements available, the better, so the acquisition of rotor information allows the extraction of information about the bearings themselves, as discussed in Section 7.7.

7.6.2.2 Applying Constraints

As shown by the SVD (discussed in Section 7.5.4), the range of the singular values of the matrix $[\Pi W]$ determines the number of parameters which can be extracted with reasonable accuracy. Equivalently, the criterion can be based on the eigenvalues of the regression matrix used for the Moore–Penrose inverse $[W^T \Pi^T \Pi W]$.

Too great a range of the eigenvalues is a source of ill-conditioning in the regression equation. Physically, this means that there is insufficient information to discriminate all the unknown parameters, and the main techniques for dealing with this issue are discussed in Section 7.5.3.

Whilst it is important to assign enough freedom to a model to obtain a good correspondence with measured data, too much freedom will result in the modelling of noise terms, and although this may improve the fit, it will reduce the predictive capability of the derived model. It is important to emphasise that the crucial test of a derived model is its predictive capability. The decision, as to the requisite number of degrees of freedom, is a difficult one, which may on occasion require some iteration. As a consequence, it is often desirable to reduce the number of unknowns in a model and this may be based on physical reasoning.

Consider the case of a machine with 14 bearings. For the sake of this argument, we consider only motion in a single direction (see the following for a discussion of the dual direction motion).

Then the stiffness matrix $\mathbf{K}$ is 14×14 and has 105 ($=14 \times 15/2$) independent terms. However, this could be reduced significantly if we assume that each bearing can only interact with its three nearest neighbours. If this is the case, then the number of terms in $\mathbf{K}$ will be $50 = (14 + 13 + 12 + 11)$, more than halving the size of the matrix. The matrix will now be of the form

$$
\mathbf{K} = \begin{bmatrix}
x & x & x & x & 0 & 0 & 0 & 0 & 0 & 0 & 0 & 0 & 0 & 0 \\
x & x & x & x & x & 0 & 0 & 0 & 0 & 0 & 0 & 0 & 0 & 0 \\
x & x & x & x & x & x & 0 & 0 & 0 & 0 & 0 & 0 & 0 & 0 \\
x & x & x & x & x & x & x & 0 & 0 & 0 & 0 & 0 & 0 & 0 \\
0 & x & x & x & x & x & x & x & 0 & 0 & 0 & 0 & 0 & 0 \\
0 & 0 & x & x & x & x & x & x & x & 0 & 0 & 0 & 0 & 0 \\
0 & 0 & 0 & x & x & x & x & x & x & x & 0 & 0 & 0 & 0 \\
0 & 0 & 0 & 0 & x & x & x & x & x & x & x & 0 & 0 & 0 \\
0 & 0 & 0 & 0 & 0 & x & x & x & x & x & x & x & 0 & 0 \\
0 & 0 & 0 & 0 & 0 & 0 & x & x & x & x & x & x & x & 0 \\
0 & 0 & 0 & 0 & 0 & 0 & 0 & x & x & x & x & x & x & x \\
0 & 0 & 0 & 0 & 0 & 0 & 0 & 0 & x & x & x & x & x & x \\
0 & 0 & 0 & 0 & 0 & 0 & 0 & 0 & 0 & x & x & x & x & x \\
0 & 0 & 0 & 0 & 0 & 0 & 0 & 0 & 0 & 0 & x & x & x & x
\end{bmatrix}
\tag{7.29}
$$

Now let us consider how to eliminate the unwanted degrees of freedom: note that this is the same as constraining them to be zero. Returning to Equation 7.2, suppose that one or more components of $\mathbf{S}$ are constrained to take specified values $\mathbf{S}_c$, then dividing the vector $\mathbf{S}$ into free and constrained elements

$$
\begin{bmatrix} \mathbf{W}_f & \mathbf{W}_c \end{bmatrix} \begin{Bmatrix} \mathbf{S}_f \\ \mathbf{S}_c \end{Bmatrix} = \{\mathbf{F}_0\}
\tag{7.30}
$$

which may be expanded to give

$$\mathbf{W}_f \mathbf{S}_f = \mathbf{F}_0 - \mathbf{W}_c \mathbf{S}_c \tag{7.31}$$

In the particular, but very common, case of constraining a parameter to zero, this is achieved simply by deleting the column of the matrix $\mathbf{W}$ corresponding to the degree of freedom in question. For non-zero constraints, the forces term is also modified as illustrated in Equation 7.31.

For the sake of illustration, consider a machine with two bearings, then in identifying full stiffness mass and damping matrices, there will be nine degrees of freedom. If the three cross-terms are to be eliminated, using the ordering convention outlined earlier, the vector of unknowns becomes

$$\mathbf{v} = \{k_{11} \quad k_{22} \quad 0 \quad m_{11} \quad m_{22} \quad 0 \quad c_{11} \quad c_{22} \quad 0\}^T \Rightarrow \{k_{11} \quad k_{22} \quad m_{11} \quad m_{22} \quad c_{11} \quad c_{22}\}^T$$

This means that columns 3, 6 and 9 of the $\mathbf{A}$ matrix are deleted, and so if the measurements cover N-frequency components, then the modified $\mathbf{A}$ matrix has the dimensions $N \times 6$. This is perhaps a rather arbitrary example, but now consider a more realistic scenario. It is reasonable to assume that a force at the high-pressure turbine end of the machine will not influence the exciter, that is there will be no significant coupling between bearing 1 and bearing 14, assuming the case of a 14-bearing turbo-generator. So the stiffness, mass and damping matrices have a banded structure, the extent of which is, *a priori*, again a matter of judgement which may be aided by considering the singular values of the measurement matrix. Recalling that the unknown terms in the vector are just the elements of the system matrices, the ordering is arbitrary and may be chosen for computational convenience.

It is less obvious how to deal with the damping in the system; there is certainly no justification to assume classical (proportional) damping in a rotating machine, and this might lead to the representation of the damping as a full matrix. However, this is not necessary and indeed it is undesirable, potentially leading to significant errors. There are two important reasons for this: firstly, the measured data of any system will only have a significant information content of the damping in a relatively restricted frequency range around the natural frequencies. Hence, identifying the damping in this way introduces a significant number of variables for which there is limited information. A second related issue for rotating machines is that, in the case of oil journal bearing mounts, about 99% of the total damping arises in the bearings, and this dissipative force is already represented within the calculated bearing forces. In principle, therefore, it would be physically reasonable to set the foundation damping to zero, but experience has shown that better numerical stability is obtained if there is a damping term within the identified model.

Although they can be identified together, again the size of the problem can be reduced by careful study of the physical behaviour of real machines. The problems of machine diagnostics are most acute in the case of large machines mounted on low-tuned (usually steel) foundations, and these machines are almost invariably supported on oil-film journal bearings. This being the case, the overwhelming proportion of damping in the system takes place in the journal oil film and consequently, the damping properties of the structure may be represented with simply a diagonal **C** matrix. It should, perhaps, be emphasised that some damping terms have been found necessary to obtain convergence in identification.

An interpretation of this is that in the experimental results being analysed, the noise level is of a similar order to those terms necessary to resolve differences between possible solutions. Figure 7.2 shows comparison of measured and calculated response on a large laboratory test-rig subjected to a known imbalance. The figure shown is typical of several from the experiments as reported by Smart et al. (2000).

Note that at all frequencies above about 15 Hz, the agreement is excellent in terms of both magnitude and phase. The reason for poor performance at low frequencies is unclear, but it is of little practical importance on real plant.

FIGURE 7.2
Comparison of measured and calculated responses. (Reproduced from Smart, M.G., et al., *Proc. R. Soc. A*, 456, 1583, 2000. With permission.)

7.6.3 Modal Approach with Kalman Filters

The alternative to the identification of matrices described earlier is the use of a modal model and this has been pioneered by several authors in Italy (Provasi et al. 2000, Vania 2000, Zanetta 1992). Once modal behaviour has been determined, then clearly all dynamic behaviours can be recovered. The big advantage of using the modal approach is a significant reduction in the number of parameters to be identified, but a penalty is paid for this insofar as the equations are non-linear in the unknown terms.

The force is inferred from the bearing properties, and hence, in contrast to the Lees approach, measurements of rotor position, in addition to pedestal motion, are required. Feng and Hahn (1995) retain physical rather than modal parameters, but in their initial study, they too use the calculated bearing coefficients with nominal static loadings to calculate the bearing dynamic forces. In more recent work, the same authors (Feng and Hahn 1998) reported the measurement of bearing oil pressure circumferentially, integrating to obtain the force exerted on the rotor. There were, understandably, significant practical difficulties in doing this. Whilst this is a very interesting laboratory investigation, it is not seen as a practical scheme for operational plant.

For several years, there has been study of this problem in Italy, involving a number of workers. Vania (2000) and Zanetta (1992) have played major roles in the work of this group. Their approach differs from that followed by the Swansea team in three respects, two of which are of physical importance, whilst the third is perhaps, more a mathematical detail. Having said that, at the time of writing, the whole problem is not understood to the extent of making categorical statements about the relative importance of different aspects.

The three aspects distinguishing the work of Zanetta and Vania et al. are as follows:

1. A modal model, rather than a physical model is used.
2. Forces are calculated using calculated bearing parameters.
3. Identification is done by Kalman filtering (discussed in Section 7.6.4).

It is the author's belief that each of these points has advantages and disadvantages, some of which raise fundamental issues as to the eventual use as to these methods. It must be considered highly likely that some approach will emerge which incorporates elements of both approaches. At present, however, the route of such a synthesis is far from clear.

In expressing the model in modal terms, two advantages are immediately introduced. Firstly, the dimensionality of the problem may be significantly reduced and secondly, the behaviour of each bearing pedestal may be written independently. The price paid for this is that the whole problem becomes very non-linear with the important implication that any feasible solution scheme must be iterative. All such methods, including the Kalman filter approach, require an initial trial state, from which the final solution can be developed. This may be a substantial disadvantage in practice, and it is well known that the absolute convergence and solution efficiency of such algorithms are strongly dependent on the closeness of the initial state to the final solution.

There is an inherent assumption of linearity in reducing the system to a modal model. At present, this may not be too restrictive an assumption, since the models used by all authors in this area are currently linear. However, retaining the model in physical space rather than modal space preserves the option of introducing non-linearities at a later stage. The price paid for this luxury is a much larger, coupled problem. It results, however, in a linear, although large problem.

The response at each location satisfies the expression

$$z_i = \sum_{r=1}^{n} \psi^i_r \frac{\psi^T_r F(\omega)}{\left(\omega_r^2 - \omega^2 + 2j\xi_r\omega_r\omega\right)} + \psi^i_l \frac{\psi^T_l F(\omega)}{-\omega^2} + \psi^i_n \frac{\psi^T_n F(\omega)}{\omega_n^2} \qquad (7.32)$$

Note that the last two terms on the right-hand side of this equation are there to allow for the influence of modes below and above the frequency range of the study. This is a device commonly used in modal analysis (see Ewins 2000). The forces are calculated from bearing properties and the differential motion. Others (e.g. Feng and Hahn 1995) have relied on a solution of Reynold's equation to achieve this. This involves some uncertainty with regard to both bearing internal clearances and static load, which, of course, can vary with operational conditions. Notice that Equation 7.32 is inherently of the output error type and hence, the conditioning will tend to be superior. It is, however, non-linear, thus introducing some mathematical complexity. The parameters to be determined are

$$x = \left\{\psi^T_1, \ldots, \psi^T_r, \ldots, \psi^T_n, \ldots, \xi_1, \ldots \xi_r, \ldots, \xi_n, \omega_1, \ldots, \omega_r, \ldots \omega_n, \psi^T_l \psi^T_n\right\}^T \qquad (7.33)$$

To give this some context, consider a machine with 14 bearings, restricting attention to vibration in a single direction. Then for n modes, the number of variables (or degrees of freedom) will be $(14 \times n + 2n + 14 \times 2)$.

Applying non-linear corrections is relatively straightforward. In the analysis presented earlier, the derived model is purely linear, but there is nothing in the fundamental approach which limits attention to a linear model. Non-linear bearing models could be introduced into this scheme, although the sensitivity of results to input uncertainties has not been assessed to date. In this regard, the input would include a detailed non-linear bearing model. This dependence on a bearing model may, however, be removed by measuring the instantaneous position of both the rotor at each journal together with the bearing pedestal motion. Indeed, these measurements are readily available on some newer turbo-generators.

At worst, the approach used in ordinary balancing procedures may be employed: All conventional balancing techniques are based on the assumption of linearity. In practice, this may lead to an iterative approach, in which successive refinements to the state of balance are made using essentially a quasi-linear approach at each stage.

Whilst shaft measurement is often found on newer turbo-generator units, many machines do not have this facility, and the provision and maintenance of the instrumentation would not be cost beneficial, or at least, it is not currently deemed to be so.

A more onerous requirement is an assessment of the static load on the bearings. This will rarely, if ever, be the case. Multi-rotor systems are aligned under cold conditions and corrections are made for anticipated thermal movements. However, there is little information on the alignment of machines under operating conditions. This is important because the dynamic characteristics of oil journal bearings are strongly dependent on the static load carried by the bearing. In some current methods, this difficulty is circumvented by the use of nominal values in the calculation. It is important, however, to appreciate that this is a source of error in the overall estimation of parameters.

7.6.4 Essentials of Kalman Filtering

Kalman filtering can be regarded as a continuous form of LS analysis which is extensively used in control applications. Here, a brief discussion is given, but further insight can be gained from numerous textbooks. A very clear explanation is given by Noton (1972) and an interesting application to structures is given by Simonian (1979), although it is now rather dated. Whereas in the standard LS method, a single calculation is made which minimises the sum of the squares of the errors terms, in the filtering approach, an initial estimate (i.e. guess!) of the answer is modified as each set of data is added. As the calculation evolves, estimates of the unknown parameters and their weightings are formed and the calculation is self-consistent insofar as the appropriate weightings are used. In this respect, filtering has an advantage over the straightforward application of the Moore–Penrose inverse calculation. In this type of calculation, each new piece of data is used in turn to give a current fit. At stage i, for example the parameter estimate is given in the form

$$y_i = \alpha_i y_{i-1} + (1 - \alpha_i) x_i \tag{7.34}$$

where x is the estimate calculated using the current set of data. In simple terms, at each stage of the calculation, the overall estimate is formed as a weighted mean of the existing estimate and that derived from the 'new' set of data. When compared with the straightforward LS approach, the filtering method does avoid the very large matrix operations by splitting the calculation into a set of rather smaller calculations. However, the calculations are far from trivial and the effective number of variables is considerable and it is worth considering the reasons for this. In the case of the stiffness, mass and damping matrices each having n degrees of freedom, the total number of variables will be $3 \times n \times (n + 1)/2$, but the weighting terms in Equation 7.13 are dependent on the variance matrices of each of the three variables. This will involve a total of $3 \times (n^2 + n) \times (n^2 + n + 1)$ coefficients; hence, these calculations are demanding in terms of storage requirements. Efficient algorithms are available and these large correlation matrices can be calculated directly from the updated parameter estimates.

7.7 Imbalance Identification

One advantage of the direct modelling approach (Smart et al. 2000) is that it is a straightforward matter to add the eccentricity vector to the list of variables for identification. This is discussed in detail by Lees and Friswell (1997), but actually, the method is a straightforward development of the method discussed in the previous section. Writing the basic equation of motion in its simplest form (following Smart et al. 2000)

$$\mathbf{Z}_F \mathbf{r}_{F,b} = \mathbf{f}_{F,b} \tag{7.35}$$

The bearing force $\mathbf{f}_{F,b}$ can be calculated as

$$\mathbf{f}_{F,b} = \mathbf{Z}_1 \mathbf{r}_{F,b} + \mathbf{Z}_2 \mathbf{f}_u \tag{7.36}$$

where

$$\mathbf{Z}_1 = \mathbf{Z}_B \left(\mathbf{P}^{-1} \mathbf{Z}_B - \mathbf{I} \right) \qquad \mathbf{Z}_2 = \mathbf{Z}_B \mathbf{P}^{-1} \mathbf{Z}_{R,bi} \mathbf{Z}_{R,oo}^{-1} \tag{7.37}$$

and

$$\mathbf{P} = \mathbf{Z}_{R,bb} + \mathbf{Z}_B - \mathbf{Z}_{R,bi} \mathbf{Z}_{R,ii}^{-1} \mathbf{Z}_{R,ib} \tag{7.38}$$

Whilst these formulae are complicated, their physical meaning is straightforward, and the matrices used involve only known or measured quantities and they are readily calculated. Using this notation, the equation can be cast in form for identification as

$$\mathbf{W}\mathbf{v} = \mathbf{f}_{F,b} = \mathbf{Z}_1 \mathbf{r}_{F,b} + \mathbf{Z}_2 \omega^2 \varepsilon \tag{7.39}$$

where ε is the unknown mass eccentricity vector (defined at each of the preselected balance planes). This equation is easily re-arranged as

$$\left[\mathbf{W} - \mathbf{Z}_2 \omega^2 \right] \begin{Bmatrix} \mathbf{v} \\ \varepsilon \end{Bmatrix} = \mathbf{Z}_1 \mathbf{r}_{F,b} \qquad (7.40)$$

The estimation of imbalance is a valuable device, and in contrast to the foundation parameters, estimates can be readily verified using balancing techniques. Table 7.1 shows the results of a series of tests on a large four-bearing laboratory rig. Six imbalance configurations were tested and the magnitude and phase of the estimates are shown.

Having established validity of this method in the laboratory, it was applied 'blindly' to a 350 MW turbo-alternator (overall length about 40 m) which had suffered a blade loss, the details of which were not revealed to the team at Swansea University. Calculations indicated high imbalance at the two balance planes on the HP rotor. This was equated to a single imbalance by considering moments and the estimates were compared with the measurements taken by the company involved. The imbalance estimate was 15% low and the estimated position along the rotor was 500 mm away from the location of the actual damage blade. This is considered very acceptable in view of the fact that the actual fractured portion of blade has an irregular shape. Some error was found in the phase estimate and this has never been resolved fully; possibly there were phase discrepancies in the signals but a full investigation of this was not possible. Nevertheless, rig results together with plant experience suggest this is a useful approach.

TABLE 7.1

Comparison of Estimated and Actual Imbalance

Test No.	Bal. Plane	Magnitude (kg m) Actual	Phase (°) Actual	Magnitude (kg m) Estimated	Phase (°) Estimated
1	3	0.160	45	0.1678	44.4
2	1	0.160	195	0.1797	188.2
	3	0.160	45	0.1606	37.5
3	3	0.160	45	0.1715	48.4
	4	0.160	150	0.2108	149.1
4	1	0.160	195	0.1737	195.9
	5	0.160	0	0.1811	−0.02
5	1	0.160	195	0.1624	182.3
	3	0.160	45	0.1318	42.9
	4	0.160	150	0.1941	154.2
6	1	0.160	195	0.1852	195.7
	3	0.160	45	0.1829	42.2
	5	0.160	0	0.1830	2.77

The method has also been to recover details of shaft bends. As discussed in Chapter 4, bends give rise to synchronous vibration, but it is important to distinguish them from imbalance. The extension to the calculation is straightforward and is not discussed here. The interested reader can find a full discussion in Edwards et al. (1998).

7.8 Extension to Misalignment

The inclusion of rotor motion data, where available, is discussed in Section 7.5 and as with all additional data, if used correctly, it can enhance the quality of the estimations and consequently, understanding of the machine's operation. However, it also opens two additional facilities to the overall technique. The first effect is very simple in physical terms; because there is, in effect, a monitor of bearing forces together with the differential motion between shaft and stator, the two may be used to assess bearing condition. The properties of the bearing, at any given configuration, depend on the static load carried and consequently, given the properties, a calculation can be set up whereby the load is determined by comparing the measured stiffness with an idealised short bearing (see Hamrock et al. 2004), but any other bearing model could be used if deemed more appropriately for a particular case. This calculation is inherently non-linear, so an output error is appropriate. The vector of unknowns is now $\theta = \{v \ \varepsilon \ F_1 \ \cdots \ F_n\}^T$. An iterative scheme is used to reach the solution. The calculation is slightly more involved than those presented in this chapter, but full details are given by Sinha et al. (2004).

7.9 Future Options

Most of the studies discussed in this chapter have emerged in the last two decades, and considerable progress has been reported on the identification of large machines and indeed, inverse problems in general. It is a prime example of an area where measured data and modelling expertise complement each other, to yield an enhanced understanding of the operation of plant. As discussed, several approaches to the problems have been presented and are now at a usable stage. Currently though, these techniques do add some complexity to the task of data analysis and there is a need for some degree of automation. Whilst this will demand some effort, it is not difficult to envisage a future system which will activate automatically on each rundown and produce estimates of the mass and stiffness terms at each bearing. In addition, the system would identify the unbalance profiles of the rotor and information on any bending, although hopefully this would be fairly rare. Extending the system to give an estimate of the misalignment could also prove invaluable in plant operation. The net result of all this would be a trend towards saving the physical parameters of a system as they develop over time, rather than the current practice of storing vast quantities of vibration data, which require considerable expert interpretation.

Looking further ahead, there are some intriguing possibilities. After a model has been derived following a rundown, it could be fine-tuned using a filtering approach. The long-term aim must be to gain maximum insight from data of machines on-load. These data are plentiful but currently of limited use, because the information content is much lower than the transient data. However, in a regime of fine-tuned models, the analyst may be able to examine the consequences of subtle on-load changes and hence gain further insight.

7.9.1 Implementation

Although this procedure might appear to some fairly complex, it is purely algorithmic and, in time, can be readily automated. This would allow significant information gain following each machine rundown. This sequence is readily followed:

1. Rundown automatically processed as described earlier.

   There is rather a modest overhead in doing this: the rotor model would need to be set up for each class of machines, that is, on a power station with, say, four machines, there will probably be a single rotor model. This point has been discussed in several papers (e.g. Lees 1988). The variations that exist between nominally identical machines are functions of their supporting structure and these differences can be treated. Ideally, rotor models should be validated using a free-free test.

2. Estimated unbalance trended against previous values (classified among hot and cold rundowns).

3. Foundation parameters compared with previous values.

   With the automated calculations, separate records could be assembled to monitor foundation parameters, state of balance, rotor bend, bearing (static) forces and hence alignment.

4. Alignment changes monitored.

5. If proximity probe data are available on the unit, the bearing characteristics may be inferred and may be indicative of wear within the bearing.

   With a separate treatment, a full record of bearing properties can be assembled. This is again made possible by the derivation of the forces at the bearings by use of a good rotor model.

Given such a system, a great deal of insight into a machine's condition can be gained with virtually no additional equipment or effort about that which is expended currently. Basic research has now developed techniques for the calculation of the parameters mentioned earlier, although there is still discussion among several research groups with regard to the most efficient methods. When these issues have been resolved, significant software development will be required to implement these approaches. However, prototype schemes could be installed in the very near future.

A further intriguing possibility arises; because after analysis, we have full knowledge of the rotor's position in the bearings and the unbalance forces, full information may be recovered using the model of the rotor to yield the deflected shape of the rotor, thus highlighting any areas of rubbing.

7.9.2 Benefits

By pursuing this analysis, what is being done is at all times seeking to fit the observed behaviour and the numerical model. Regarding the model and the plant data as sources of information, a judicious combination of the two will yield the maximum information about the state of the plant. This idea may be surprising to some who might argue that the plant data are 'real', whilst the model is a mere figment of imagination. This, however, is not the case. Any piece of information which enhances plant understanding is to be valued. Information from plant and model should be viewed as complementary evidence as to the overall state and can operate as a cross-check on each other.

Problems

7.1 Table 7.2 shows the response of a two-spring, two-mass system in the range of 5–100 Hz. The system is excited by a unit force acting on mass 2. Formulate the regression equation to determine the stiffness mass and damping parameters.

7.2 Repeat the calculation of Problem 7.1, using only data up to a frequency of 50 Hz. Explain the difference between the derived estimates and those obtained using the complete frequency range.

7.3 Using the data in Table 7.2, estimate the system parameters K, M and C, assuming that the damping matrix is diagonal. Compare the estimated models displacement with those measured in the table.

7.4 Calculate the singular values of the regression matrix of the data shown in Table 7.2 and show how this varies as data are added to the system.

TABLE 7.2

Input Data for Problems 7.1 through 7.4

Frequency (Hz)	Displacement 1		Displacement 2	
	Magnitude (μm)	Phase (°)	Magnitude (μm)	Phase (°)
10	1.00	−1	2.43	−1
20	1.15	−11	2.71	−8
30	1.74	−28	3.79	−24
40	3.09	−87	5.84	−80
50	1.74	−149	2.59	−138
60	1.13	−168	1.11	−150
70	1.1	179	0.46	−136
80	2.19	144	1.13	−67
90	0.86	28	1.13	−161
100	0.27	12	0.61	−170

7.5 A uniform rotor of diameter 350 mm has an overall length of 3 m. At one end is a fan which may be represented as a steel disc of diameter 600 mm and thickness 150 mm.

The two bearings are at the drive end of the shaft and 2 m along the rotor. The imbalance on the fan is 0.0003 kg m. At full speed, the shaft motion at the bearing locations is measured as

	x-Direction	y-Direction
Bearing 1	6×10^{-6} m at 30°	4×10^{-6} m at −20°
Bearing 2	8×10^{-6} m at 20°	5×10^{-6} m at 70°

Determine the dynamic bearing forces.

References

Childs, D., 1993, *Turbomachinery Rotordynamics: Phenomena, Modeling and Analysis*, Wiley, New York.

Edwards S., Lees, A.W. and Friswell, M.I., 1998, The identification of a rotor bend from vibration measurements, in *16th International Modal Analysis Conference*, Santa-Barbara, CA, pp. 1543–1549.

Ewins, D.I., 2000, *Modal Testing: Theory, Practice and Application*, 2nd edn., Wiley-Blackwell, U.K.

Feng, N.S. and Hahn, E.J., 1995, Including foundation effects on the vibration behaviour of rotating machinery, *Mechanical Systems and Signal Processing*, 1995, 9(3), 243–256.

Feng, N.S. and Hahn, E.J., 1998, Vibration analysis of statically indeterminate rotors with hydrodynamic bearings, *Journal of Tribology – Transactions of ASME*, 120(4), 781–788.

Friswell, M.I., Penny, J.E.T., Garvey, S.D. and Lees, A.W., 2010, *Dynamics of Rotating Machines*, Cambridge University Press, New York.

Golub, G.H. and Van Loan, C.F., 2012, *Matrix Computations*, The John Hopkins University Press, Baltimore, MD.

Hamrock, B.J., Schmid, S.R. and Jacobson, B.O., 2004, *Fundamentals of Fluid Film Lubrication*, Marcel-Dekker, NJ.

Hirdaris, S.E. and Lees, A.W., 2005, A conforming unified finite element formulation for the vibration of thick beam and frames, *International Journal for Numerical Methods in Engineering*, 62(4), 579–599.

Jeffcott, H.H., 1919, The lateral vibration of loaded shafts in the neighbourhood of a whirling speed: The effects of want of balance, *Philosophical Magazine*, 6, 37, 304–314.

Kramer, E., 1993, *Dynamics of Rotors and Foundations*, Springer-Verlag, Berlin, Germany.

Lees, A.W., 1988, The least square method applied to identify rotor foundation parameters, in *Institution of Mechanical Engineers Conference on Vibrations in Rotating Machinery*, Edinburgh, U.K., pp. 209–215.

Lees, A.W. and Friswell, M.I., 1997, The evaluation of rotor imbalance in flexibly mounted machines, *Journal of Sound and Vibration*, 208(5), 671–683.

Lees, A.W. and Friswell, M.I., 2001, The vibration signature of chordal cracks in asymmetric rotors, in *19th International Modal Analysis Conference*, Kissimmee, FL.

Lees, A.W. and Simpson, I.C., 1983, Dynamics of turbo-alternator foundations, in *Institution of Mechanical Engineers Conference*, London, U.K., Paper C6/83.

Mayes, I.W. and Davies, W.G.R., 1976, The behaviour of a rotating shaft system containing a transverse crack, *Institution of Mechanical Engineers Conference on Vibrations in Rotating Machinery*, Cambridge, U.K., pp. 53–64.

Nelson, H.D., 1980, A finite rotating shaft element using Timoshenko beam theory, *Journal of Strain Analysis for Engineering Design*, 102, 793–803.

Nelson, H.D. and McVaugh, J.M., 1976, The dynamics of rotor-bearing systems using finite elements, *Journal of Engineering for Industry*, 98, 593–599.

Noton, M., 1972, *Modern Control Engineering*, Pergamon Press, Inc., Elmsford, NY.

Oldfield, M., Ouyang, H. and Mottershead, J.E., 2005, Simplified models of bolted joints under harmonic loading, *Computers & Structures*, 84(1–2), 25–33.

Pennacchi, P., Bachschmid, N., Vania, A. et al., 2006, Use of modal representation for the supporting structure in model-based fault identification of large rotating machinery: part 1 – theoretical remarks, *Mechanical Systems and Signal Processing*, 20(3), 662–681.

Pons, A., 1986, Experimental and numerical analysis on a large nuclear steam turbo-generator, in *IFToMM Conference on Rotordynamics*, Tokyo, Japan, p. 269.

Provasi, R., Zanetta, G.A. and Vania, A., 2000, The extended Kalman filter in the frequency domain for the identification of mechanical structure excited by multiple sinusoidal inputs, *Mechanical Systems and Signal Processing*, 14(3), 327–341.

Simonian, S.S., 1979, System identification in structural dynamics: An application to wind force estimation, PhD thesis, University of California, University Microfilms International, Ann Arbor, MI.

Sinha, J.K., Friswell, M.I. and Lees, A.W., 2002, The identification of the unbalance and the foundation model of a flexible rotating machine from a single run-down, *Mechanical Systems and Signal Processing*, 16, 255–171.

Sinha, J.K., Lees, A.W. and Friswell, M.I., 2004, Estimating the static load on the fluid bearings of a flexible machine from run-down data, *Mechanical Systems and Signal Processing*, 18, 1349–1368.

Smart, M.G., Friswell, M.I. and Lees, A.W., 2000, Estimating turbogenerator foundation parameters – Model selection and regularisation, *Proceedings of the Royal Society A*, 456, 1583–1607.

Stodola, A., 1927, *Steam and Gas Turbines*, Translation by S. Loewenstein, McGraw-Hill Book Co., Inc., New York.

Vania, A., 2000, On the identification of a large turbogenerator unit by the analysis of transient vibrations, in *Institution of Mechanical Engineers Conference on Vibrations in Rotating Machinery*, Nottingham, U.K., pp. 313–322.

Worden, K. and Tomlinson, G.R., 2000, *Non-linearity in Structural Dynamics*, CRC Press, Boca Raton, FL (Originally published by IoP Press, Bristol, PA.)

Zanetta, G.A., 1992, Identification methods in the dynamics of turbogenerator rotors, in *Institution of Mechanical Engineers Conference on Vibrations in Rotating Machinery*, Bath, U.K., paper C432/092, pp. 173–182.

8

Some Further Analysis Methods

8.1 Standard Approaches

There are several distinct aspects to the data processing involved in assessing the condition of a piece of machinery ranging from purely empirical to more physics-based approaches. Workers in the field tend to concentrate on certain aspects within this broad range of activities, but few would doubt the need for this broad ranging approach to the problems.

The overall aim is to extract information from data. Of itself, a vibration reading is of little value unless translated into context and hence given some physical meaning but the ways of doing this form the subject of this chapter. This conversion of data into information is by no means trivial and warrants some study. At one end of the spectrum would be a purely statistical approach. For instance a user may build up a large bank of vibration and other machine parameters and suggest that any marked change in the pattern signifies some change in the machine which should then be investigated. This is indeed a simple model, but it may be appropriate to some relatively low-value plant where the user may specify that a given change in some features of the parameter should trigger removal of the machine concerned from service.

In dealing with vibration data, appropriate functions might be the root-mean-square vibration or other functions such as the Kurtosis, which basically measures the spikiness of a signal and gives a good indication of impacts within the system. Slightly more complex systems would involve measuring vibration in two planes and recording on a scatter diagram as discussed in Chapter 2. As shown there, some statistical tests are readily introduced to assess the point at which a change of significance has taken place. Of course, the data recorded in such a system will usually have more than just vibration data but will also incorporate records of speed, load and performance.

Whilst this simplest level system may be adequate to guide the safe operation of plant, for more complex (and valuable) machines, the requirements will often be rather more sophisticated. On the more critical items of plant,

rather more is required from a monitoring system. It is common in the literature to classify monitoring schemes into three categories as follows:

1. Detection of a fault
2. Diagnosis and localisation of a fault
3. Correction or mitigation of the effects of a fault

Each of these stages includes preceding steps. Category 1 has already been
discussed and the crucial requirement at this level is just some criteria to
alert an alarm or shutdown signal. This can be done on purely statistical
grounds, but insight may be required to identify which parameter should be
used as a guide to the condition. As described earlier, one very simple measure is the RMS value of vibration. Let $y(t)$ be the measured vibration, then

$$y_{\text{rms}}(t) = \sqrt{\frac{1}{T} \int_{t-\frac{T}{2}}^{t+\frac{T}{2}} y^2(t)dt} \qquad (8.1)$$

where the period T is chosen to suit the problem in hand. For machines
which operate under nominally stationary conditions, a fairly long time window can be used, whereas a machine in transient conditions will require a
rather narrower time window. This is a useful concept even in cases where
there are multiple components of vibration components, but it may not fully
reflect the important features of a problem. For instance, consider the monitoring of a rolling-element bearing. Of particular interest is the presence and
magnitudes of impacts between different components of the bearing. The
Kurtosis is a measure of the 'spikiness' of a signal and hence can be helpful
in highlighting important changes. The Kurtosis is defined as

$$K(t) = \frac{\int_{t-(T/2)}^{t+(T/2)} y^4(t)dt}{\left(\int_{t-(T/2)}^{t+(T/2)} y^2(t)dt\right)^2} \qquad (8.2)$$

where T is selected with the same considerations as outlined earlier.
However, the use of this measure as a monitoring parameter relies on some,
albeit limited, insight into the operation of the machine involved.

Figure 8.1 shows a signal which is predominantly Gaussian random noise.
The issue here is how the engineer can unequivocally discriminate this from
a signal comprising several sinusoidal components. In fact, there are a number of approaches to this, but the examination of Kurtosis is possibly the
most convenient.

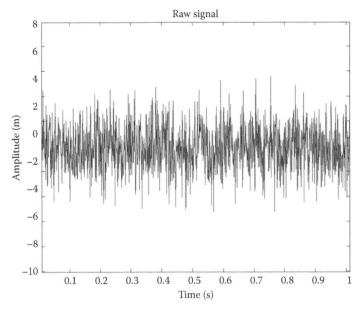

FIGURE 8.1
Noise.

Other purely statistical approaches include the use of the auto and cross-correlation and their respective spectral representations. These methods are useful in cases which have a significant random component to the measurement as might be the case on a fluid handling machine, for example. For a signal with a substantial random element, a common approach to dealing with this is to calculate the autocorrelation of the signal which in physical terms expresses how often the signal repeats itself.

Given a measurement $y(t)$ the autocorrelation function is calculated as

$$R(\tau) = \lim_{T \to \infty} \frac{1}{T} \int_{0}^{T} y(t)y(t + \tau)dt \tag{8.3}$$

In practice, of course, T will be finite, but the longer it is, the better will be the averaging out of random terms to reveal the underlying structure of the signal. This pre-supposes that the signal is stationary: although there are substantial fluctuations, the statistical properties of y do not vary with time. Note that R is a function only of the interval between the readings and not of the absolute time.

Let us look at these in an example in which the signal is heavily contaminated with noise. Part of the raw time signal is shown in Figure 8.1.

The figure shows the first second of the raw data, and it is immediately apparent that any analytic form is heavily submerged in noise. Because there

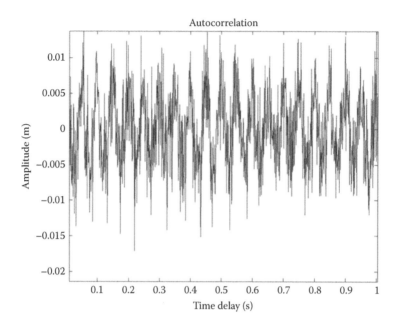

FIGURE 8.2
Autocorrelation function.

is so much noise, considerable averaging is required to obtain the true signal. In Figure 8.2, the autocorrelation is shown after averaging over 1000 s:
of course, there is a substantial assumption here that the system is constant
for such a length of time – a condition that is rarely satisfied in rotating
machinery. This is discussed in Section 8.6.6. For the present discussion, it is
assumed that a suitable period of steady operation is available. The variation
of the autocorrelation is shown over a period of 1 s and when compared with
Figure 8.1, it is immediately obvious how the periodic structure emerges.
Note that although there is still significant noise, underlying patterns are
clearer.

The extent of noise introduced in this example far exceeds that which
would normally be observed in machinery, but it has been exaggerated to
illustrate the point. Although Figure 8.2 is still noisy, the structure is clear,
and its Fourier transform, the power spectral density, is shown in Figure 8.3.
The two frequency components in the original signal are immediately
apparent at 20 and 40 Hz, respectively.

In this case, these two components can also be identified by performing an
FFT directly on the original signal, but the noise level is rather higher and
the component at 40 Hz becomes partially submerged. This is clear from
Figure 8.4, which shows the FFT for a period of 1000 s.

The progression through these stages has gradually clarified the underlying structure of what, at first, appeared to be a purely random signal. It is
difficult to present a precise recipe for processing data in this way, but there

FIGURE 8.3
The power spectral density.

FIGURE 8.4
FFT over 1000 s of data.

are standard tools at the disposal of the analyst, some of which have been applied in this case. Whilst some of the standard techniques will often be applied, some prior knowledge of the operation of the machine producing the data in question is often a distinct advantage.

The important point to note is that all the basic signal-processing approaches are purely statistical in nature. The basic role of the analyst is to seek to infer a physical interpretation from the available data. This will often involve a combination of techniques, the choice of which will be dependent on the nature of the investigation.

8.2 Artificial Neural Networks

The majority of the discussion presented has been deterministic in nature. Deterministic, physically based representations are extremely helpful because they offer the prospect of real physical insight into the operation of the machine. But this is not always possible as a number of factors which influence the operation of machines are beyond current modelling capabilities and in these instances, different approaches are needed. For instance, Chapter 2 discusses the presentation of scatter plots which show the variation of vibration levels (both magnitude and phase) over a considerable period of time. The example shown in Figure 2.6 illustrates how the vibration varies over a period of time even though the machine is running at nominally constant speed, and operators are led to ask why the machine behaves in this way.

At one level, the answer is obvious: although the machine is operating under nominally constant conditions, numerous parameters are constantly changing, but the control system is making corrections in order to hold the output constant. For instance although load and speed are (nominally) constant, the rotor current, steam pressure and condenser levels are constantly varying. Consider the case of a turbine/generator comprising six bearings. In addition to the two components of vibration at each bearing together with the shaft speed, 54 additional parameters were monitored, mainly comprising temperatures and pressures at various points around the circuit. Whilst these parameters were ideally constant, they all show some fluctuation. These fluctuations have an effect on the vibration levels but in a rather complicated way and in order to give an accurate assessment of a machine's condition, a method is needed to assess these variations. Since this is beyond the capability of current direct modelling methods, a more empirical approach is required. One such route is the use of artificial neural networks (ANNs).

An ANN is a method by which a set of input is associated with a set of output by adjusting a set of weight parameters. In the present example, the input comprises the 54 parameters of the machine (implying that in the network, there will be 54 nodes in the input layer). The output layer in this instance will have 12 nodes yielding the vibration levels at each bearing (or, more generally, each measurement point) which may be expressed as either magnitude and phase or real and imaginary parts. Between the input and output layers, there will be one or more intermediate layers and the number of nodes in each of these layers is dependent on the problem. The situation is reminiscent of many identification problems: too few nodes will give an inadequate fit, whilst too many will lead to the modelling of noise. The 54 input parameters are the various operational parameters such as pressures, temperatures, rotor current, etc. whilst the 12 output parameters will be the (complex) bearing vibration levels.

All of the nodes in each layer are linked to each node in the next layer and there is a weighting associated with each link. These weights are adjusted to give optimum fit between the inputs and outputs. This process is called 'training'. The trained network can then be used to predict the vibration given any set of parameters, provided of course, that there has been no underlying change in the machine's condition.

This may be regarded as a form of interpolation. The most common form of ANN used in condition monitoring applications is the multi-layer perceptron and its structure is shown in Figure 8.5.

The figure shows just one hidden layer, but in principle, there may be an arbitrary number. As shown in Figure 8.5, each node is connected to each

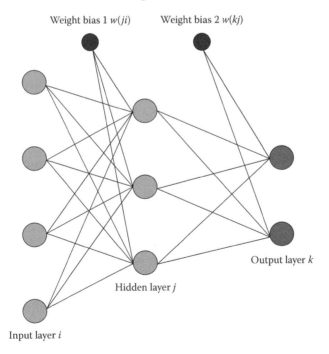

FIGURE 8.5
The multi-layer perceptron.

node of the preceding layer and it passes an output to each node of the subsequent layer. The forward signal at each stage is given by

$$y = f\left(\sum w_i x_i + w_0\right) \tag{8.4}$$

The function f can take several forms, but a common choice is the sigmoid, that is

$$f(x) = \frac{1}{1 + e^{-x}} \tag{8.5}$$

This is an appropriate function, because it has a continuous derivative which vanishes at high positive and negative values. The weighting functions w are obtained by optimising the agreement between input and output for the set of training data. Because the function f (called the activation function) is non-linear, it can be shown that any set of data may be represented by the network.

There are, of course, some pitfalls in the application of these techniques. Because the inputs may cover a wide variety of physical measurements ranging from displacements of order 10^{-6} m to temperatures of hundreds of degrees, it is important to normalise all inputs to a range of –1 to +1, to ensure that all measurements are given due significance in the calculation. This is easily achieved by simply re-scaling each input node on a scale –1 to 1. The other difficulty is the choice of network parameters to use; too few hidden layers will restrict the ability to fit to data, but an excessive number will result in the modelling of noise. This dilemma is encountered in many methods of system identification.

Note that as this fit is obtained by adjusting the weights, this process may be considered as an advanced form of curve fitting and this may be regarded as both its strength and weakness: using an appropriate network represents a convenient means of obtaining some functional representation of a set of data, but this is just a curve fit and does not necessarily relate to any physical parameters of the system, in marked contrast to the work presented in Chapter 7. A full discussion of the training of ANNs is beyond the scope of this chapter and the interested reader is referred to Worden and Tomlinson (2001), for example. There are a number of user-friendly routines available for the application and manipulation of ANNs, see, for example Nabney (2003).

Of greater importance to this chapter is a discussion of the usefulness in the context of machine health monitoring. Being a sophisticated form of curve fitting, information may be interpolated, but not extrapolated. This is simple, because the data are simply fitted rather than associated with any underlying physical equation. However, used appropriately, an ANN can help give considerable physical insight into machine behaviour. We illustrate this point with the following example associated with data in the scatter diagrams of Chapter 2.

The data shown in the scatter diagram of Figure 2.6 are plots of the vibration at one of the bearings over a period of 11 days. There are 8600 individual measurements, and each of these depends on 54 parameters such as rotor speed, steam temperature and pressure together with a range of other measurements which although nominally constant, do vary due to continual adjustments in the control system to meet varying load conditions. The overall task is to determine how the bearing temperature depends on each of the parameters. There is no reason to presume that this dependence is linear, but we can gain some understanding of some important issues by initially assuming a linear relationship. Although one might suspect that not all of the 54 variables are of equal importance, there is a need to establish an approach to assess the relative importance of the variables: in other words, which terms are driving the observations.

Let θ represent the temperature at each time, P be the array of parameters and R contain the unknown coefficients. Then

$$\theta = PR \qquad (8.6)$$

In this equation, for a single bearing (or other measurement location), θ is a vector of length 8600 in this example, $\mathbf{P}$ is a matrix 8600 × 54 and R is a vector of length 54. The Moore–Penrose inverse may be applied to this problem to give

$$R = \left(\mathbf{P}^{\mathsf{T}}\mathbf{P}\right)^{-1}\mathbf{P}^{\mathsf{T}}\theta \qquad (8.7)$$

Unfortunately, this answer will be unreliable as it is subject to considerable noise. To shed some light on this problem, we modify the approach and analyse the matrix $\mathbf{P}$ using singular value decomposition (SVD) (see Section 8.6.1). It is known that

$$\mathbf{P} = \mathbf{U}\Sigma\mathbf{V}^{\mathsf{T}} \qquad (8.8)$$

Σ is a diagonal matrix containing the 54 singular values in non-ascending order: $\mathbf{U}$ and $\mathbf{V}$ are matrices with ortho-normal columns and $\mathbf{U}^{\mathsf{T}}\mathbf{U} = \mathbf{V}^{\mathsf{T}}\mathbf{V} = \mathbf{I}_n$. There are several algorithms available to achieve this decomposition (see, e.g., Golub and van Loan 2012), including a standard function in MATLAB. The mechanics of these algorithms need not be covered here: more important is its usefulness. Manipulating Equation 8.6 yields

$$\Sigma^{-1}\mathbf{U}^{\mathsf{T}}\theta = \mathbf{V}^{\mathsf{T}}R \qquad (8.9)$$

A formal solution for R may be obtained by expressing it in terms of the columns of $\mathbf{V}$; hence,

$$R = \sum_{j=1}^{54} a_j v_j \tag{8.10}$$

From this, we obtain

$$a_j = \frac{u_j^T \theta}{\sigma_j} \tag{8.11}$$

It is apparent here that terms for which σ_j is small make a significant difference to the calculated answer, but this is of concern for reasons which can be appreciated by returning to Equation 8.6. Combining Equations 8.8 and 8.10 shows that

$$\mathbf{P}v_j = \sigma_j u_j \tag{8.12}$$

This means that there are some vectors which simply produce noise in the system, and the way to avoid this noise is simply to truncate the range of singular values which are considered in the solution, in effect setting all the rest to zeros. Actually, this type of analysis can lead to considerable insight. This will now be illustrated by returning to the example at hand.

FIGURE 8.6
Singular values of bearing data.

The two graphs shown in Figure 8.6 are only marginally different. They differ only insofar as some clearly erroneous readings have been removed from the 'modified' data. The figure shows the distribution of singular values, and it is very clear that the rate of changes with respect to index (column) number changes markedly around the value $x = 12$, but clearly the precise location is somewhat debatable. At higher indices, owing to the low singular values, higher terms make little contribution yet; if included, these terms would severely distort the solution. Hence, terms in these higher modes are regarded as noise and ignored.

8.3 Merging ANNs with Physics-Based Models

ANNs are non-linear methods which can provide an excellent set of tools for fitting observational data. In this respect, they are much more powerful than linear regression, but this 'flexibility' can also be a disadvantage in some circumstances. One benefit of using a regression technique is that the engineer can fit data to a given equation form and thereby impose some constraints on the underlying physics. There would be significant advantages to find some form of network, which could constrain the fit in this manner. Although this concept may appear somewhat counter-intuitive, it can be an advantage to seek a less optimum fit. In Chapter 7, the identification of some foundation models is improved by choosing the appropriate degrees of freedom to evaluate. Typically, it is found that fitting to a large number of degrees of freedom will invariably improve the 'fit', but the derived model has little or no predictive capability. The same argument applies to ANNs: constraining the form of the derived model may lead to improved capability, but it is far from obvious how this might be achieved directly. Hopefully, some method of applying these ideas will emerge in due course.

There are, however, indirect ways in which the accuracy and flexibility of ANNs can be merged with regression approaches. In Section 8.2, we discuss how the influence of pseudo-static parameters can influence the behaviour and ANNs can be used to separate the effects of different parameters. Once this has been achieved, a separate regressive form can be developed.

8.4 Kernel Density Estimation

One of the objectives of this chapter is the presentation of a range of analysis techniques which either have been proved useful or, in the case of newer developments, are considered promising.

Such an approach is kernel density estimation which may be used for the estimation of probability distribution functions from measured data. Given a limited number of measurements of a variable x_j the problem is to estimate the probability distribution, $p(x)$. Of course, if there were an infinite number of measurements, this would be a trivial exercise, but in reality, some subtle approach is required particularly if there is more than a single variable. The first step is fairly obvious – one simply divides the range into a series of bins. Mathematically this may be expressed as

$$p(x) = \frac{1}{n} \sum_{j=1}^{n} \frac{1}{h} w\left(\frac{x - x_j}{h}\right)$$

(8.13)

where
   n is the number of data points
   h is the window width
   x_j is the sampled data and

$$w(r) = \frac{1}{2} \quad |r| < 1$$

$$w(r) = 0 \quad |r| \geq 1$$

(8.14)

Whilst this gives some idea of the distribution, it is rather crude and is discontinuous. This is a particular problem where the number of samples, n, is small in some sense. This distribution is sometimes called the naïve estimate. Basically, this has been constructed by placing a 'box' of width h and the resulting so-called naïve estimate is shown in Figure 8.7.

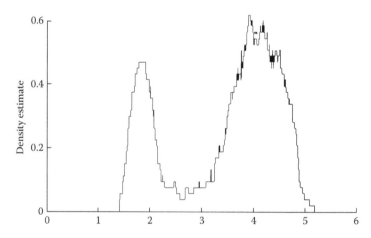

FIGURE 8.7
Naïve estimate.

An approximate probability distribution can be formed by taking n over-lapping bins of width $2h$ and height $(2nh)^{-1}$ in the appropriate position for each measurement and then summing. It is implicitly assumed that each measured value is at the centre of a bin.

The situation can be substantially improved simply by inserting a Gaussian curve instead of a rectangular box for each measurement point, or 'atom' as they are (rather confusingly) called. The improvement stems from the fact that the Gaussian and its derivative is well behaved in the sense of being continuous and single valued in contrast to the top hat function. Hence, we can form an estimate as

$$\hat{f}(X) = \frac{1}{n}\sum_{i=1}^{n}\frac{1}{h}K\left(\frac{X-x_i}{h}\right)$$ (8.15)

where K can, in principle, be any well-behaved distribution, but the Gaussian is normally used so that

$$K(x) = \frac{1}{\sqrt{2\pi}}\exp\left(-\frac{x^2}{2}\right)$$ (8.16)

Inserting the Gaussian into Equation 8.15, the only parameter to be chosen is h, the so-called bandwidth. This is chosen to minimise the mean integrated square error (between the derived distribution and the actual readings). Silverman (1986) describes how this is used to derive the optimal choice for the bandwidth and the result is

$$h = 0.9An^{-1/5}$$ (8.17)

Where A is min(standard deviation, inter-quartile range/1.34). In the litera-ture, there are several (minor) variations in the magnitude of this expression. In principle, the precise optimum will depend on the shape of the (unknown) distribution. One can, of course, formulate an iterative procedure, but in real-ity, errors from this source will be fairly small.

The way in which smooth curves (Gaussian being the most common) are used as components to form distribution curves is illustrated in Figure 8.8. The choice of h, the bandwidth, or sometimes called the 'smooth-ing parameter' is, of course, crucial. The effect of changing this parame-ter is shown in Figure 8.8. This figure shows the importance of choosing the appropriate bandwidth to get insight into the underlying probability distribution.

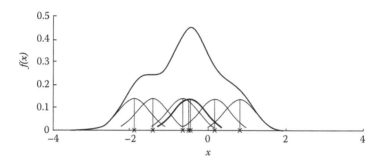

FIGURE 8.8
Composite distribution curve.

The situation becomes slightly more complicated where there are two or more variables. In the case of two variables, the estimated density function is given by

$$\hat{f}(v) = \frac{1}{n}|H|^{-1/2} \sum_{i=1}^{n} K\left(H^{-1/2}[v - X_i]\right) \tag{8.18}$$

where

$$H = \begin{bmatrix} h_1^2 & h_3 \\ h_3 & h_2^2 \end{bmatrix} \tag{8.19}$$

h_1 and h_2 are the smoothing parameters for the x and y values, respectively, and h_3 controls the orientation. It is assumed that

$$h_3 = \rho h_1 h_2 \tag{8.20}$$

with $0 < \rho < 1$. As one might anticipate, the situation becomes rapidly more complicated for higher dimensions.

So, how can these techniques be used to advantage in condition monitoring? Whilst only limited use has been made to date, it would appear that there is significant potential. Baydar et al. (2001) reported the application of KDE techniques in the study of gearbox wear. It was shown that the improved representation of the probability distribution could be used to detect an incipient fault at an earlier stage than was possible with more conventional approaches.

This reasoning would also suggest that this approach can also benefit studies in remaining life prediction of machine components. The author, however, is not aware of any work to date in this area.

8.5 Rapid Transients: Vold–Kalman Method

In Chapter 2, the analysis of machine transient operation is discussed and it is shown how the study of run-up and rundowns often yields significant insight into a machine's condition. In other words, it is information rich. There is a contradiction inherent in the analysis of transient phenomena with essentially steady-state methods based on Fourier methods. As discussed in Chapter 2, this places a limitation on the frequency resolution which can be obtained and, perhaps not surprisingly, the resolution decreases as the speed of the transient increases. For machines which undergo rapid transients, an alternative approach is needed and a suitable methodology was developed in the 1990s.

The approach was reported by Vold and Leuridan (1995) although there have been a number of developments since then. The method does not involve Fourier transforms, thereby avoiding some of the inherent difficulties of resolution. Nevertheless, the approach does involve very significant computing demands. To begin the analysis, we observe that the vibration signal may be expressed as

$$x(t) = \sum A_k(t)\sin(k\omega t + \phi_k) + \eta(t) \tag{8.21}$$

Note that the shaft speed ω is itself a function of time, and η is the so-called nuisance component: this includes that part of the signal which is not related to the shaft speed as well as any noise contributions. The task now is to design a filter which has input $x(t)$ and output $A_k(t)$. Of course, this requires some constraint on the variation of the signal which gives the so-called structural equation.

In the original formulation of the method, it was assumed that the signal follows a sinusoidal variation and hence, using standard trigonometric identities

$$\sin(\omega k \Delta t) + \sin(\omega[k-2]\Delta t) = 2\cos(\omega t)\sin(\omega[k-1]\Delta t) \tag{8.22}$$

Hence, the structural equation may be written as

$$x(n\Delta t) - 2\cos(\omega \Delta t)x([n-1]\Delta t) + x([n-2]\Delta t) = \varepsilon(t) \tag{8.23}$$

where the term on the right-hand side is recognition of the fact that the signal is not strictly sinusoidal. Before proceeding with the analysis, it is worth considering Equation 8.23 a little further: note that the speed term ω is varying in time, a fact which is the whole basis of our problem and hence, a more precise expression for the structural equation would be

$$x(n\Delta t) - 2\cos\left(\int_0^{[n-1]\Delta t} \omega(t)\,dt\right) x([n-1]\Delta t) + x([n-2]\Delta t) = \varepsilon(t) \quad (8.24)$$

The total angular rotation at the nth time step may be written as

$$\Theta(n\Delta t) = \int_0^{n\Delta t} \omega(t)\,dt \quad (8.25)$$

Using this notation, the measured data at the nth point may be expressed as

$$y(n) = x(n)e^{j\Theta(n)} + \eta(n) \quad (8.26)$$

where $x(n)$ represents the (complex) amplitude of the wave $e^{j\Theta(n)}$, whilst $\eta(n)$ is a 'noise' term. Note that for normal diagnostics, $x(n)$ is the term of interest. Equation 8.26 is conveniently re-cast in the form

$$\{\eta\} = \{y\} - [C]\{x\} \quad (8.27)$$

where

$$[C] = \begin{bmatrix} e^{j\Theta(1)} & 0 & \vdots & 0 \\ 0 & e^{j\Theta(2)} & \vdots & 0 \\ \cdots & \cdots & \ddots & \cdots \\ 0 & 0 & \vdots & e^{j\Theta(n)} \end{bmatrix}$$

The square of the error vector norm is given by

$$\{\eta\}^H\{\eta\} = (\{y\}^T - \{x\}^H[C]^H)(\{y\} - [C]\{x\}) \quad (8.28)$$

However, we are not yet in a position to complete the analysis of the transient data. To achieve this goal, some assessment must be given as to how rapidly the wave amplitude can change and this is done via the so called structural equation. For example, using a two pole filter representation gives

$$x(n) - 2x(n+1) + x(n+2) = \varepsilon(n) \quad (8.29)$$

Noting that the use of a higher-order filter is equivalent to allowing more rapid changes in the response carrier amplitude, Equation 8.29 may be conveniently written as

$$[A]\{x\} = \{\varepsilon\} \tag{8.30}$$

For the two-pole case,

$$\begin{bmatrix} 1 & -2 & 1 & 0 & \cdots & 0 & 0 & 0 \\ 0 & 1 & -2 & 1 & \cdots & 0 & 0 & 0 \\ \cdots & \cdots & \cdots & \cdots & \cdots & \cdots & \cdots & \cdots \\ 0 & 0 & 0 & 0 & \cdots & 1 & -2 & 1 \end{bmatrix} \begin{bmatrix} x(1) \\ x(2) \\ \cdots \\ x(N) \end{bmatrix} = \begin{Bmatrix} \varepsilon(1) \\ \varepsilon(2) \\ \cdots \\ \varepsilon(N-2) \end{Bmatrix} \tag{8.31}$$

The square of the error term can now be expressed as

$$\{\varepsilon\}^T \{\varepsilon\} = \{x\}^T [A]^T [A]\{x\} \tag{8.32}$$

Hence, in this approach, there are two distinct error terms to be minimised, and the weighting, r, between these two largely dictates the nature of the solution obtained. The first term refers to the magnitude of measurement noise, whilst the second determines the rate at which the response envelope can change. The global solution is obtained by minimising

$$J = r^2 \{\varepsilon\}^T \{\varepsilon\} + \{\eta\}^T \{\eta\} \tag{8.33}$$

Expanding this expression gives

$$J = r^2 \{x\}^T [A]^T [A]\{x\} + \left(\{y\}^T - \{x\}^H [C]^H \right) \left(\{y\} - [C]\{x\} \right) \tag{8.34}$$

This weighted penalty function can be easily minimised as

$$\frac{\partial J}{\partial x^H} = \left(r^2 [A]^T [A] + [E] \right) \{x\} - [C]^H \{y\} = 0 \tag{8.35}$$

This leads to a solution

$$\{x\} = \left(r^2 [A]^T [A] + [E] \right)^{-1} [C]^H \{y\} \tag{8.36}$$

where

$$[E] = [C]^H [C] \tag{8.37}$$

It is important to appreciate that this approach circumvents the difficulty of resolution which is inherent in Fourier approaches by fitting directly to harmonics of the shaft speed and it is in this respect, it has something in common with Prony analysis. The fundamental limitation on the 'accuracy' of the approach arises from the assumption made on the maximum rate of change of the vibration levels as reflected in the choice of r, the weighting factor in Equation 8.36. This is actually quite a complicated issue which depends on the rate at which the shaft is accelerated or decelerated and the system is damping.

As a machine changes speed, the vibration levels at any one time will not equate to the levels corresponding to the instantaneous speed because the system does not have time to saturate. Because of this, the example shown here is somewhat idealistic insofar as immediate response to forcing is assumed: this does have the virtue, however, of providing a known target for our calculations. Figure 8.9 shows the ideal frequency response function of our test rotor but, in this case, no dependence on the weighting factor.

For a more realistic model, the transient profile will be dependent on the rate at which the speed is changed. This can be seen from a comparison of the observed vibration levels as our machine is run down from 3000 rev/min in 100, 10 and 2 s, respectively. These are shown in Figures 8.10 through 8.12.

FIGURE 8.9
The exact frequency response.

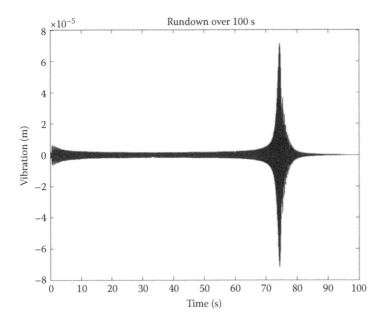

FIGURE 8.10
100 s rundown.

FIGURE 8.11
Faster rundown.

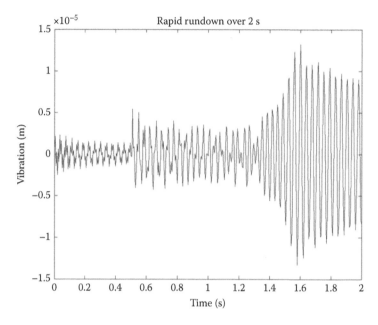

FIGURE 8.12
Very rapid rundown.

It is clear that the 100 and 10 s rundowns bear a similarity to each other, showing substantial excitation as the machine passes through the lower resonance. However, the rundown over 2 s is radically different, and it becomes difficult to identify the underlying dynamic behaviour.

This is because the system speed is changing so rapidly that the system cannot reach its steady-state value at each speed. Perhaps more importantly, these rapid transients cannot be adequately analysed using the short-time Fourier transform (STFT) as described in Chapter 2. Table 8.1 shows the maximum resolution that can be obtained by STFT for the three rates of rundown considered.

In Table 8.1, the parameter α is the exponential speed decay rate, that is $\Omega = \Omega_0 e^{-\alpha t}$.

TABLE 8.1

Resolution for Exponential Speed Reduction

Rundown time (s)	α	Resolution (Hz)
2	1.152	7.59
10	0.23	3.39
100	0.023	1.07
1000	0.0023	0.34

Of course, for uniform sampling, the resolution will improve as the speed decreases, but these figures give the resolution at the upper end of the speed range. Consider now the application of the Vold–Kalman approach on the rapid rundown giving a resolution of 0.34 Hz (the same as for the slowest rundown). Given the sampling rate used, the Nyquist frequency is 100 Hz. Tůma (2005) gives an expression relating the weighting factors used: for a two-pole filter

$$r = \sqrt{\frac{\sqrt{2}-2}{6-8\cos(\pi\Delta f)+2\cos(2\pi\Delta f)}} \approx \frac{0.00652097315}{\Delta f^2} \tag{8.38}$$

In this case, $\Delta f = 0.34/100$ and hence $r = 564$. This yields the appropriate frequency resolution, but the price paid for this is reduced smoothing of signal noise. Without detailed prior knowledge of the system, this compromise cannot be avoided. Figure 8.13 shows the synchronous excitation calculated with weighting terms of 564, 100 and 50, respectively, using the rapid rundown data.

The most rapid transient shown here, in Figure 8.12, is at first sight confusing and no clear evidence of the critical speeds or indeed shaft orders are immediately apparent. It is clear from Table 8.1 that only very poor

FIGURE 8.13
Comparison of weighting effects.

Variation of response with weighting

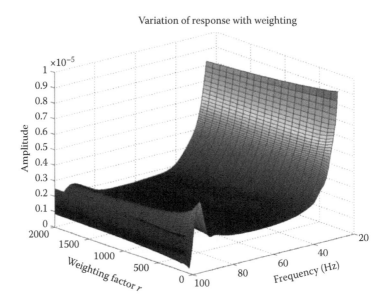

FIGURE 8.14
Variation of the Vold–Kalman method.

resolution can be obtained using FFT techniques. Nevertheless, application of the Vold–Kalman approach can yield valuable information.

Figure 8.14 shows a waterfall plot in which the synchronous vibration component is plotted against shaft speed and weight factor and it is immediately apparent that two critical speeds have been influenced in this speed range.

It is worth noting that the shape of the response derived is dependent on the value chosen for the weighting factor. As discussed in the preceding pages, the weighting factor expresses the ratio of the errors in the amplitude variation and the signal noise in the overall 'target' which is optimised. Therefore, a low value of the weighting permits rapid transients in amplitude of response, whereas it is clear that at higher values, the peak at about 90 Hz is considerably suppressed. In this case, a second-order filter has been applied.

At this point, it is important to note some remarks concerning the amplitude of the responses close to resonance. Although it is often of only minor importance in problem resolution, in the case of slow transients, the response to a given imbalance is dependent on mode shape and damping only. With rapid transients, however, the rate of change of rotor speed has also a marked influence on the peak response. This is because any system takes some time to reach its full steady-state response to sinusoidal excitation, a fact which is frequently used to advantage in the deliberate rapid transit through critical speeds. This is an important topic and the interested reader is referred to the discussion given by Friswell et al. (2010).

8.6 Useful Techniques

8.6.1 Singular Value Decomposition

In discussions elsewhere (see Sections 8.2 and 7.5.4), singular value decomposition has been applied, but in view of its usefulness, a brief account of the technique is appropriate. SVD is a direct extension of eigenvalue analysis to matrices which are not square. Consider a rectangular matrix $\mathbf{A}$ with dimensions $m \times n$. The case $m > n$ is of particular interest as this corresponds to an over-specified problem for which some approach such as least squares must be applied to reach a satisfactory solution. All matrices $\mathbf{A}$ may be decomposed as the product of three matrices

$$\mathbf{A} = \mathbf{U\Lambda V} \tag{8.39}$$

where $\mathbf{U}$ has dimensions $m \times m$, $\mathbf{V}$ has dimensions $n \times n$ and Λ has the dimensions of $\mathbf{A}$. The matrices $\mathbf{U}$ and $\mathbf{V}$ are orthogonal, that is

$$\mathbf{U}^T\mathbf{U} = \mathbf{I}_m \quad \mathbf{V}^T\mathbf{V} = \mathbf{I}_n \tag{8.40}$$

where $\mathbf{I}_n$ is the unit (diagonal) matrix of dimension n. The matrix Λ is given by

$$\Lambda = \begin{bmatrix} \lambda_1 & 0 & 0 & 0 \\ 0 & \lambda_2 & 0 & 0 \\ 0 & 0 & \ddots & 0 \\ 0 & 0 & 0 & \lambda_n \\ 0 & 0 & 0 & 0 \\ 0 & 0 & 0 & 0 \end{bmatrix} \tag{8.41}$$

From Equations 8.40 and 8.41, two equations may be developed to yield

$$\mathbf{AV}^T = \mathbf{U\Lambda} \quad \mathbf{U}^T\mathbf{A} = \Lambda\mathbf{V} \tag{8.42}$$

Note here that if $\mathbf{A}$ is square, then $\mathbf{U}$ and $\mathbf{V}$ are identical and Equation 8.42 becomes the standard eigenvalue problem.

Now consider the meaning of all this using an over-specified problem of the form

$$\mathbf{As} = \mathbf{f} \tag{8.43}$$

This can be re-expressed as

$$U \Lambda V s = f \tag{8.44}$$

Therefore, using the orthogonality relationships

$$\Lambda V s = U^T f \tag{8.45}$$

If we now examine the pth column of matrix U, then if the corresponding singular values (λ_p) are low (relative to the rest of the set), this means that the combination of parameters $(U^T f)$ contributes little to the overall equation, but the corollary of this is that this low sensitivity can lead to significant errors. The way around this is a truncation of the series of eigenvalues as used in Section 7.4.4. It is then a straightforward matter to invert Equation 8.45. Although the matrix Λ is rectangular, an inverse can be established because of its particular form.

Often this indicates that the model has too many parameters for the data available and this is a useful check in many investigations. The SVD gives good insight into the structure of a data set. A full account of the mathematical details is given in a number of textbooks (e.g., Golub and van Loan 2012). From a user's standpoint, SVD is a very easy means of analysis via, for example MATLAB.

8.6.2 The Hilbert Transform

Whilst the Fourier and Laplace transforms project a function into a different domain (e.g., time to frequency), the Hilbert transform (HT) remains in the same space and hence, there is a fundamental distinction between this and some of the other common transformations. However, the HT gives some deep insights into the nature of a signal. Here, only a brief discussion is given on the nature, and main uses of the transform are described. Readers with more specific requirements are referred to the more extensive discussions by Worden and Tomlinson (2001) or Feldman (2011).

The definition of the transform is

$$H[x(t)] = \tilde{x}(t) = \pi^{-1} \int_{-\infty}^{\infty} \frac{x(\tau)}{t - \tau} d\tau \tag{8.46}$$

This straightforward definition in itself gives little insight into the application of the transform. It has the effect of 'smearing out' the time variation of the function. The full meaning of the transform is in fact rather deep, and there are a number of ways of approaching it. In the parlance of signal processing, HT may be regarded as a linear filter in which all amplitudes remain unchanged, yet the phases are shifted by $-\pi/2$. The transform has a range of properties, the details of which can be found in a number of sources (see, e.g. Feldman 2011, Worden and Tomlinson 2001).

Owing to the difficulty in forming the integral due to the singularity, it is taken as the Cauchy integral or principal value.

The focus here, however, is how HT can be useful in the analysis of signals from machinery. The two most important applications are the derivation of a signal envelope and the detection of non-linearity: both of these depend on the so-called analytic signal, again a function of time but now complex. This is defined as

$$X(t) = x(t) + j\tilde{x}(t) \tag{8.47}$$

where $\tilde{x}(t)$ is the HT of $x(t)$.

Treating this as any other complex function, one may immediately write

$$A(X) = \sqrt{x^2(t) + \tilde{x}^2(t)} \tag{8.48}$$

Now consider as an example the decaying sine wave. Let

$$x(t) = Ae^{-at} \sin \omega t \tag{8.49}$$

Then,

$$H(x(t)) = \tilde{x}(t) = Ae^{-at} \cos \omega t \tag{8.50}$$

Therefore,

$$A(t) = |X(t)| = Ae^{-at}\left(\sqrt{\sin^2 \omega t + \cos^2 \omega t}\right) = Ae^{-at} \tag{8.51}$$

Hence, a simple application of the transform provides a means of extracting the envelope of a signal. This is often the most salient feature of a signal rather than the more rapid variations that can mask overall features. In addition to the envelope amplitude, Equation 8.51 also yields a value for α, and the time derivative of this yields the instantaneous frequency.

Individual HTs can be proved tricky to calculate: the good news is that tables exist so that full calculations are rarely, if ever, needed in practice. As an example, consider the important function $\cos \omega t$. For the HT

$$H\{\cos \omega t\} = \frac{-1}{\pi} P \int\limits_{-\infty}^{\infty} \frac{\cos \omega(y+t)}{y} \, dy \tag{8.52}$$

where the last expression is obtained simply by changing variable. This is now expanded to give

$$H\{\cos \omega t\} = \frac{-1}{\pi} \left\{ \cos \omega t P \int_{-\infty}^{\infty} \frac{\cos \omega y}{y} \, dy - \sin \omega t P \int_{-\infty}^{\infty} \frac{\sin \omega y}{y} \, dy \right\} \qquad (8.53)$$

The first integrand on the right-hand side is an odd function and hence, the integral becomes zero. The evaluation of the second term is somewhat more involved. It may be evaluated by considering the contour integral

$$I = \oint \frac{e^{j \omega z}}{z} \, dz \qquad (8.54)$$

with $\omega > 0$ and the contour encircles the upper half plane, but circumvents the origin. The overall integral is divided into four sections corresponding to two parts of the real axis, then two semicircle sections, one infinitely large and one infinitesimally small. By Cauchy's theorem, it is known that the total contour integral is zero and from this, it can be deduced that

$$\int_{-\infty}^{\infty} \frac{\sin \omega y}{y} \, dy = \pi \qquad (8.55)$$

Using this, it follows that

$$H\{\cos \omega t\} = \sin \omega t \qquad (8.56)$$

An important feature of signals from mechanical systems is causality. A system cannot respond until an excitation has been applied to it. For the purpose of this discussion, it suffices to define a causal function as one which has zero value for negative time, and clearly, this is a pre-requisite for any description of a physical system. An elegant discussion of this issue is given by Worden and Tomlinson (2001) and only a brief summary is given here.

For any function of time $g(t)$, there is a decomposition

$$g(t) = g_{\text{even}}(t) + g_{\text{odd}}(t) \qquad (8.57)$$

But if $g(t)$ is causal,

$$g_{\text{even}}(t) = \frac{g(|t|)}{2}, \quad \text{all } t \qquad (8.58)$$

And

$$g_{odd}(t) = \frac{g(|t|)}{2} \quad t > 0$$

$$= -\frac{g(|t|)}{2} \quad t < 0 \tag{8.59}$$

But, as observed by Worden and Tomlinson (2001), these two equations may be written more elegantly (and meaningfully!) as

$$g_{even}(t) = g_{odd}(t)\varepsilon(t)$$

$$g_{odd}(t) = g_{even}(t)\varepsilon(t) \tag{8.60}$$

where $\varepsilon(t)$ is the signum function (often denoted sgn(t)) given by

$$\varepsilon(t) = \begin{cases} 1 & t > 0 \\ 0 & t = 0 \\ -1 & t < 0 \end{cases} \tag{8.61}$$

Noting that the Fourier transform of this function is $j/\pi\omega$, the relationship with the HT emerges and Equation 8.46 may be re-expressed as

$$\text{Re}(G(\Omega)) = -\frac{1}{\pi} \int_{-\infty}^{\infty} \frac{\text{Im}(G(\Omega))}{\Omega - \omega} d\Omega \tag{8.62}$$

$$\text{Im}(G(\Omega)) = \frac{1}{\pi} \int_{-\infty}^{\infty} \frac{\text{Re}(G(\Omega))}{\Omega - \omega} d\Omega \tag{8.63}$$

This gives a succinct but very profound conclusion: the Fourier transform of a causal function is invariant under HT, that is,

$$G(\omega) = \tilde{G}(\omega) = H\{G(\omega)\} \tag{8.64}$$

These relationships can also be utilised to assess the linearity (or otherwise) of a signal.

In summary, the HT has two major roles in the area of machine monitoring:

1. The determination of envelope curves
2. The detection of non-linearity

8.6.3 Time Frequency and Wigner–Ville

Section 2.2.4 discusses the standard method of analysing data from a machine during transient operation. This is important, because, although it presents a number of analysis challenges, the data obtained during transients are rich in information as discussed in Chapter 7. In Chapter 2, the problem was resolved by imposing a window in time which moves as the transient progresses. The general term for this approach is the STFT and this proves a very useful tool despite some difficulties. The problem arises from the difficulty of defining frequency and time simultaneously, and this is not merely a mathematical issue but rather a matter of examining the underlying concept. The frequency of a sinusoid is perfectly defined, but a sine wave is not limited in time. On the other hand, a time signal that is localised in time contains a spread of frequencies. In the case of the STFT, the transformation of the signal can be written as

$$S(t, \omega) = \int x(\tau) w(t - \tau) e^{j\omega\tau} d\tau \tag{8.65}$$

where the window function is given by

$$w(t) = 1 \quad 0 < t < \delta$$

$$w(t) = 0 \quad \text{otherwise}$$

where δ is the width of the window. As discussed is Chapter 2, the choice of this width is a compromise between the resolution of the FFT and the acceleration/deceleration of the rotor.

Insight into the development of the spectrum in time can be gained by analogy with the way in which stationary signals may be analysed. A useful measure is the autocorrelation function of a stationary signal that is given by

$$R(\tau) = \int x(t) x(t + \tau) dt \tag{8.66}$$

This will reveal the periodicities by showing the relationship of two signals separated by a time τ. For non-stationary signals, the analogous quantity will also be a function of time and hence, this cannot be integrated. So, we may write

$$R(t, \tau) = x^* \left(t - \frac{\tau}{2} \right) x \left(t + \frac{\tau}{2} \right) \tag{8.67}$$

where suffix * denotes complex conjugate.

And the Wigner–Ville distribution is the Fourier transform of this quantity, and this gives a view of the spectral content of the signal as a function of time:

$$S_w(t,\omega) = \int x^* \left(t - \frac{\tau}{2} \right) x \left(t + \frac{\tau}{2} \right) e^{-j\omega\tau} d\tau \tag{8.68}$$

It may be observed that the expression for the STFT shown in Equation 8.67 is actually an approximation – its precise form is dictated by the window function w. Whilst the STFT has some intuitive credibility, there is some lack of precision in the inability to define the instantaneous frequency.

The issues raised in this discussion are very relevant to several subsequent points. Indeed the concept of time and frequency is central to a number of issues in machine monitoring. The formalism expressed here is helpful in outlining other section.

8.6.4 Wavelet Analysis

It is true to say that the Fourier transform and its digital descendants (DFT, FFT) form the backbone of signal processing as it is applied to rotating machinery monitoring and diagnostics. It is indeed a convenient approach and has the immense mathematical advantage that it converts a differential equation of motion into an algebraic equation. Provided that the system is linear, or at least predominantly so, then this proves an extremely convenient approach, so much so that it has become embedded in the engineer's way of thinking; frequency is thought of as an independent entity. All this is very attractive in many instances, and yet there are fundamental conflicts when transient conditions are to be considered. By its very nature, a sinusoid acts over all time; any pulse or transient behaviour must be represented as a combination of frequency components and this leads to significant complication. It is the case that most methods for the analysis of machinery are based on the assumption of stationarity. For example, in examining the transient operation of turbo machines, it is common to consider them to be nearly stationary using the so-called STFT. Indeed, it is conceptually difficult to image precisely what is meant by a frequency of say 50 Hz, if it persists for only a very short period of time.

This brings the discussion into the realm of time-frequency problems, an area with its own very substantial literature. Amongst a wide class of techniques, one of the most prominent is the wavelet transformation. Conceptually, the distinction between a Fourier resolution and one involving wavelets is that in the latter, the basic unit is localised in time unlike the sinusoid function. A common wavelet function is the Morlet function given by

$$\Psi(t) = c_\sigma \pi^{-1/4} e^{-t^2/2} \left(e^{i\sigma t} - e^{-1/2\sigma^2} \right) \tag{8.69}$$

where c_g is for normalisation. There are several forms of wavelet in common usage, but the essential point is that they are functions which are localised in time. A full discussion of the requirements of a wavelet is beyond our present scope, but the interested reader may refer to the wide ranging review of methods by Feng et al. (2013).

Given a signal $x(t)$, whereas the STFT proves a useful tool in many cases, the discussion of preceding sections illustrates some of the shortcomings, all related to the time/frequency conflict. One possible way of treating transient problems effectively is by means of the wavelet transformation. This takes the form

$$W_x(a,b;\psi) = a^{-1/2} \int x(t)\psi^* \left(\frac{t-b}{a} \right) dt \qquad (8.70)$$

where * denotes the complex conjugate. ψ is the wavelet chosen and so W is a transformation of x involving two variables a and b in just the same way as the STFT is a transform involving two variables, in that case t and ω.

A wide range of functions may be chosen as wavelets, essentially they are pulses which are limited in time the Morlet function is just one of the common choices. The two main requirements are that the wavelet has zero mean

$$\int \psi(s)ds = 0 \qquad (8.71)$$

and the so-called admissibility condition

$$\int \frac{\psi^2(s)}{|s|} ds < \infty \qquad (8.72)$$

Wavelet decomposition offers advantages over STFT analysis and has been applied to many cases by researchers. As yet, however, it has had limited impact on the industrial community. This relatively slow take can probably be ascribed to two causes: (a) the slightly bewildering choice as to wavelet form and (b) confusion over the physical interpretation of results. Computation is not a problem as commercial packages (e.g. LABVIEW) have well-developed facilities.

It is the author's view that interpretation is the main obstacle to progress; in STFT, the meaning of time and frequency is clear despite the difficulties outlined in the previous section. In contrast, the wavelet transform yields parameters a, b. The first represents scale, whilst b is a dilation term, but the physical meaning is less clear. These issues are discussed by Peng et al. (2005) who provide an extensive literature on the application of wavelets to machine diagnostics.

8.6.5 Cepstrum

Many practitioners regard Cepstrum as a 'novel' technique even though it dates back to the early 1960s and pre-dates the FFT. One of the co-authors of the paper which introduced Cepstrum was J.W. Tukey who 2 years later was co-author of the FFT algorithm (Cooley and Tukey 1965). This first paper (Bogert et al. 1963) defines the Cepstrum as 'the power spectrum of the logarithm of the power spectrum'. The history of the technique has been admirably described by Randall (2013).

In mathematical terms, the Cepstrum is given by

$$C_p(\tau) = F^{-1}(\log(F_{xx}(\omega))) \tag{8.73}$$

where $F_{xx}(f)$ is the power spectrum, whilst F^{-1} denotes the inverse Fourier transform. Despite this apparently contorted definition, the logarithmic form is extremely useful. In the study of gears, for instance the logarithmic scale reveals sideband structure which is often not apparent in the direct calculation on a linear scale. The Cepstrum gives insight into sideband structure, and this has been proved particularly useful in the analysis of gearing problems. Other notable studies have analysed steam turbines with damaged blades.

Another important feature of the Cepstrum is that, because it is logarithmic, the force terms and the impulse response properties are additive rather than multiplicative, and this can be used to simplify some analyses substantially. Hence, in general, we may write

$$x(t) = \int_0^t p(t')h(t-t')dt' = p(t) * h(t) \tag{8.74}$$

where $p(t)$ is the applied force and $h(t)$ is the impulse response, and $*$ denotes the convolution operation. Taking the Fourier transform and using the convolution theorem gives

$$X(\omega) = P(\omega) \times H(\omega) \tag{8.75}$$

where X, P and H are the respective Fourier transforms of x, p and h, respectively. Hence, taking logs

$$\log(X(\omega)) = \log(P(\omega)) + \log(H(\omega)) \tag{8.76}$$

This means an effective separation of force and system terms, a feature of considerable help in identifying sources.

Although most engineers working in the area of machine condition monitoring are aware of Cepstrum, it remains something of a speciality. It is an extremely powerful tool that, perhaps, warrants wider usage.

8.6.6 Cyclostationary Methods

The development of cyclostationary analysis is rather more recent than many methods in current use. The approach is a sophisticated signal analysis to the study of systems which are not exactly periodic, yet exhibit some cyclic behaviour. A good example of this, indeed a problem to which it has been successfully applied, is the motion of rolling-element bearings. Naïvely one might expect that such a bearing would show truly periodic behaviour, but this is not the case for a variety of reasons. Random slipping between the elements, possible speed fluctuations and variations of axial and radial loads impose slight variations on the cycle time. Although these variations may be only slight, Antoni (2007) shows that such a variation can destroy the harmonic nature of the response.

The problem is resolved by observing that the variations in frequency are statistical, and the variance of the speed fluctuations is periodic, albeit at an, as yet unknown frequency. Denoting the Fourier transform of $x(t)$ as $X(\omega)$

$$S_x(\omega;\alpha) = \lim_{L \to \infty} \frac{1}{L} E\left\{ X\left(\omega + \frac{\alpha}{2}\right) \times X^*\left(\omega - \frac{\alpha}{2}\right) \right\} \tag{8.77}$$

where L is the interval over which the signal is sampled. $S_x(\omega; \alpha)$ is called the cyclic power spectrum. Maxima of this quantity indicate the frequency content of the original signal.

This technique involves some sophisticated signal processing, but applied to appropriate problems has been proved capable of revealing spectral features not apparent using more conventional approaches. Once the precise frequencies have been established in this way, any modulation caused by the variation from true periodicity can be corrected.

8.6.7 Higher-Order Spectra

A variety of techniques are described in this text, most of which treat the vibration spectra in various ways in order to detect and/or diagnose the machine's condition. As shown elsewhere, in many cases, substantial progress can be made by assuming that the system under study is linear even though it is known that this is rarely the case in any strict sense. Indeed, it can often be a valuable step in diagnosis to attempt an evaluation of the extent of non-linear effects, and there are several ways of approaching this, one being the use of so-called higher-order spectra (HOS). In examining HOS, the analyst seeks to evaluate not only the distribution of frequency components (as in most studies), but also the relationships between them as this can yield considerable insight.

As a starting point, consider the power spectral density of a time series signal (t). This is computed as

$$S_{xx}(f_k) = \frac{1}{n}\sum_{r=1}^{n} X_r(f_k)X_r^*(f_k) \tag{8.78}$$

and gives a measure of the energy in a frequency component. This notion may be generalised to give the bi-spectrum which indicates the relationship between two frequency components, and their sum, for instance, f_k, f_l and f_{k+l}. The bi-spectrum is defined mathematically as

$$B(f_k, f_l) = \frac{1}{n}\sum_{r=1}^{n} X_r(f_k)X_r(f_l)X_r^*(f_{k+l}) \tag{8.79}$$

A bicoherence may also be defined (Fackrell et al. 1995) as

$$b^2(f_k, f_l) = \frac{\left|\sum_{r=1}^{n} X_r(f_k)X_r(f_l)X_r^*(f_{k+l})\right|^2}{\sum_{r=1}^{n}\left|X_r(f_k)X_r(f_l)\right|^2 \sum_{r=1}^{n}\left|X_r(f_{k+l})\right|^2} \tag{8.80}$$

The bicoherence at any frequency pair f_k, f_l can be interpreted as the fraction of power at F_{K+1} that is phase-coupled to the two components.

Of course, this just defines the first term of a set of HOS, but to some extent, the first few of the set are the most valuable. The next in the series is the tri-spectrum and this is given by

$$T(f_k, f_l, f_m) = \frac{1}{n}\sum_{r=1}^{n} X_r(f_k)X_r(f_l)X_r(f_m)X_r^*(f_{k+l+m}) \tag{8.81}$$

For the effective use of these tools, there are a range of considerations with regard to sampling rate, sample size and windowing techniques that are discussed by Fackrell et al. (1995). Here, attention is focussed on why these techniques are useful. A complete discussion would be ambitious indeed, but it is relatively straightforward to give an indication based on physical principles. Collis et al. (1998) also give some useful insight.

A complete hierarchy of moments can be defined so that

$$\mu_1 = \int_0^\infty x(t)dt \tag{8.82}$$

This, of course, corresponds to the mean displacement (normally zero). The next in the series gives the signal's power as

$$\mu_2 = \int_0^\infty x^2(t)dt \tag{8.83}$$

whilst the third term gives a measure of skewness. If the distribution of displacements is Gaussian, then this too will be zero:

$$\mu_3 = \int_0^\infty x^3(t)dt \tag{8.84}$$

whilst the fourth moment is just the Kurtosis:

$$\mu_4 = \int_0^\infty x^4(t)dt \tag{8.85}$$

Consider a typical dynamic equation describing the behaviour of a machine. In matrix formulation, this may be written as

$$Kx + K_n x^2 + C\dot{x} + M\ddot{x} = f(t) \tag{8.86}$$

Here at non-linear stiffness term has been included K_n. There is a slight difficulty here in rigorously defining the meaning of the cubed term in a vector sense, but it can be taken as individual components for the purposes of the present discussion. Now assume that the forcing term comprises two components with angular speeds ω_1, ω_2 and suppose that, in some sense, the non-linear terms are small, then the problem of solving Equation 8.86 may be approached by first obtaining a linear solution and then using this to apply a correction. In this way, an iterative procedure can be developed, but only the first-order correction is required for the current argument. Now let $x = x_L + x_n$, for the linear

$$x_L = \left[K + j\omega_1 C - \omega_1^2 M\right]^{-1} f_1 e^{j\omega_1 t} + \left[K + j\omega_2 C - \omega_2^2 M\right]^{-1} f_2 e^{j\omega_2 t} \tag{8.87}$$

which may be written as

$$x_L = Ae^{j\omega_1 t} + Be^{j\omega_2 t} \tag{8.88}$$

The second-order term in the solution can now be calculated using Equation 8.87 to give

$$Kx_n + C\dot{x}_n + M\ddot{x}_n = -K_n x_L^2 \tag{8.89}$$

The essence of the argument is that the forcing terms of the non-linear correction will contain terms $\left(Ae^{j\omega_1 t} + Be^{j\omega_2 t} \right)^2$ which may be expanded to give $\alpha e^{2j\omega_1 t} + \beta e^{2j\omega_2 t} + \gamma e^{j(\omega_1+\omega_2)t}$, and so it is clear that components of the response at frequency $\omega_1 + \omega_2$ are produced as a direct consequence of the system non-linearity. This can be utilised as a test for system non-linearity, and in many instances, this in itself is a good condition indicator.

The simple physical result is that if a linear system is excited simultaneously with forces at frequencies f_1 and f_2, then there will be excitation at these two frequencies and only these two frequencies. However, if the system has some non-linearity, there will also be response at $(f_1 \pm f_2)$, hence this provides an indicator of non-linear behaviour. Similarly, excitation at a single frequency will produce a range of harmonics. This rather complicated behaviour can be presented graphically by plotting the bi-spectrum against the frequency ranges. The bi-spectrum is given by

$$B_{xxx} = E[X(f_l)X(f_m)X^*(f_{l+m})] \quad f_l + f_m \le f_N \tag{8.90}$$

where

$$X(f) = 2\pi \int_0^T x(t)e^{j2\pi f} df$$

Extending the approach further, it can be helpful to consider the tri-spectra, defined as

$$T_{xxxx} = E[X(f_k)X(f_l)X(f_m)X^*(f_{k+l+m})] \quad f_k + f_l + f_m \le f_N \tag{8.91}$$

Yanussa-Kaltungo and Sinha (2014) have discussed the use of bi-spectrum and tri-spectrum for the diagnosis of faults in rotating machine and simulated faults on a laboratory rig. They studied misalignment, shaft cracks and shaft rub in their tests and different patterns emerge for each category. This is physically reasonable, since the type of non-linearity is somewhat different for each case. Their summary results suggest considerable potential for condition monitoring. The paper emphasises the need for both bi-spectrum and tri-spectrum results to distinguish between different faults. Further work is needed both in terms of experimental work and analytic studies on the types of non-linearity arising in different faults on a range of machines. Nevertheless, this is a developing area with some potential.

8.6.8 Empirical Mode Decomposition

A number of sections in the chapter discuss aspect of the analysis of non-stationary and non-linear data, and no discussion of this topic would be complete without inclusion of the empirical mode decomposition, also known as the Huang–Hilbert transform. The technique was first published in 1998 (Huang et al. 1998) and has been applied widely for the analysis of non-stationary and non-linear data. Data from machinery are, of course, often stationary, being dominated by rotational speed effects, but there are instances in which full non-stationary analysis is an advantage.

The term 'mode' in the name of the technique is somewhat deceptive in so far as it has nothing to do with the analytic mode shapes corresponding to eigenvectors of matrix models. The 'modes' used in this technique are just shapes which prove convenient for the data decomposition.

The technique can be applied to any signal, but it is only advantageous in the case of data which are non-stationary and/or non-linear. The approach is to resolve the complete data set into a linear combination of so-called intrinsic mode functions (IMFs). The real relevance of this is that these intrinsic mode functions–all have well-defined HTs, and this leads to the ability to define instantaneous energy and frequency, since the data are expressed as a sum of IMFs.

An IMF can be any function which satisfies two conditions:

1. In the whole data set, the number of extrema and the number of zero crossings must be either equal or differ by one.
2. At any point, the mean value of the envelope defined by local maxima and that defined by local minima is zero.

The decomposition is largely intuitive and is derived through an iterative process, the steps of which are as follows.

Let the input data be $x(t)$.

1. All maxima are identified and joined with an envelope (e_{max}) using a cubic spline fit.
2. All minima are identified and joined with an envelope (e_{min}) using a cubic spline fit.
3. The mean of these two envelopes is calculated at every point as $m(t) = \dfrac{e_{max}(t) + e_{min}(t)}{2}$.
4. If $m(t) = 0$, the function is an IMF; otherwise, subtract this mean function from the input and repeat the process.

5. This cycle is repeated until, ideally (4) is satisfied or some tolerance is met. The first IMF, say IMF1, is defined.

6. Subtract this function from the original data and repeat the whole process for $r(t) = x(t) - \text{IMF1}$.

7. The whole process is repeated until some error criterion is met.

For each IMF, an HT will give the instantaneous amplitude and frequency.

The real power of this approach is in its treatment of non-stationary data. It is still relatively new when compared to other techniques such as FFT but clearly has a role to play. It has been applied to some investigations on sub-harmonic vibration in gas blowers (Wu and Qu 2009) and to some interesting gearbox diagnostics (Ma et al. 2015).

8.7 Concluding Remarks

A broad range of techniques are presented in this chapter. The varying methods will be useful in different circumstances, and hence, there is no golden rule on which approach to use when. The suitability will depend on the type of data and the features to be extracted.

Problems

8.1 A uniform steel rotor, 1 m in length, is supported at each end on bearings, each of which has constant stiffness of 5×10^5 N/m and damping of 150 Ns/m. There are two discs mounted 250 mm from either end and both of these are steel with thickness 20 mm and diameter 100 mm. The first disc has an imbalance of 0.04 kg mm, whilst the second has an imbalance of 0.02 kg mm, both at zero phase. The shaft passes through a clearance of 10 μm at mid-span. Calculate the response and determine the speed at which the shaft will contact the stator.

8.2 A machine runs down from 500 rev/min in about 2 min. Suggest a means of determining the rundown curve and calculate the optimum resolution.

8.3 A machine can be represented at a uniform rotor 0.3 m diameter and 4 m between bearings, each of which has constant stiffness of 5×10^7 N/m and damping of 12,000 Ns/m. There are two discs 1 m inboard of each bearing and these two discs have 0.5 m diameter and 0.05 m thickness.

The first disc has an imbalance of 0.0003 kg m. Determine the response as a function of speed during a slow run-up.

8.4 The rotor in Problem 8.3 is accelerated steadily from rest to 3000 rev/min in 30 s. Calculate the response and compare with the observed behaviour on a slow run-up. Using the results apply the Vold–Kalman technique to obtain a rundown profile.

8.5 A time signal has the form

$$y(t) = \left(5e^{-0.05t}\sin(100\pi t) + 3e^{-0.08t}\cos(60\pi t)\right)e^{-(t-25)^2/100}$$

Determine the maximum instantaneous frequency and velocity, and the time at which these occur.

References

Antoni, J., 2007, Cyclic spectral analysis of rolling-element bearing signals: Facts and fictions, *Journal of Sound and Vibration*, 304, 497–529.

Baydar, N., Chen, Q., Ball, A. and Kruger, U., 2001, Detection of incipient tooth defect in helical gears using multivariate statistics, *Mechanical Systems and Signal Processing*, 15(2), 303–321.

Bogert, B.P., Healey, M.J.R. and Tukey, J.W., 1963, The quefrequency analysis of time series for echoes: Cepstrum, pseudo-autocovariance, cross-cepstrum and saphe cracking, *Proceeding of the symposium on Time Series Analysis*, Wiley, NY, pp. 209–243.

Collis, W.B., White, P.R. and Hammond, J.K., 1998, Higher order spectra: The bi-spectrum and tri-spectrum, *Mechanical Systems and Signal Processing*, 12(3), 375–394.

Cooley, J.W. and Tukey, J.W., 1965, An algorithm for the machine calculation of complex Fourier series, *Mathematics of Computation*, 19(90), 297–301.

Fackrell, J.W.A., White, P.R., Hammond, J.K. and Pinnington, R.J., 1995, The interpretation of the bispectra of vibration signals – 1. Theory, *Mechanical Systems and Signal Processing*, 9(3), 257–266.

Feldman, M., 2011, Hilbert transform in vibration analysis, *Mechanical Systems and Signal Processing*, 25, 735–802.

Feng, Z., Liang, M. and Chu, F., 2013, Recent advances in time-frequency analysis methods for machinery fault diagnosis: A review with application examples, *Mechanical Systems and Signal Processing*, 38, 165–205.

Friswell, M.I., Penny, J.E.T., Garvey, S.D. and Lees, A.W., 2010, *Dynamics of Rotating Machines*, Cambridge University Press, New York.

Golub, G.H. and van Loan, C.F., 2012, *Matrix Computations*, The John Hopkins University Press.

Huang, N.E., Shen, Z., Long, S.R., Wu, M.C., Shi, H.H., Zheng, Q., Yen, N.-C., Tunf, C.C. and Liu, H.H., 1998, The empirical mode decomposition and the Hilbert Spectrum for nonlinear and non-stationary time series analysis, *Proceedings of the Royal Society A*, 454, 903–995.

Ma, H., Pang, X., Feng, R., Song, R. and Wen, B., 2015, Fault feature analysis of cracked gear considering the effects of the extended tooth contact, *Engineering Failure Analysis*, 48, 105–120.

Nabney, I.T., 2003, *Netlab, Algorithms for Pattern Recognition*, Springer, London, U.K.

Peng, Z.K., Tse, P.W. and Chu, F.L., 2005, A comparison study of improved Hilbert-Huang transform and wavelet transform: Application to fault diagnosis for rolling bearing, *Mechanical Systems and Signal Processing*, 19, 974–988.

Randall, R.B., 2013, A history of Cepstrum analysis and its application to mechanical problems, *Conference Surveillance 7*, 29–30, October 2013, Chartres, France.

Silverman, B.W., 1986, *Density Estimation for Statistics and Data Analysis*, Chapman and Hall, London.

Tuma, R., 2005, Setting the passband width in the Vold_Kalman order tracking filter, *12th International Congress on Sound and Vibration*, 11–14 July, Lisbon, Portugal.

Vold, H. and Leuridan, J., 1995, High resolution order tracking at extreme slew rates using Kalman tracking filters, *Shock and Vibration*, 2, 507–515.

Worden, K. and Tomlinson, G.R., 2001, *Non-linearity in Structural Dynamics*, CRC Press, Boca Raton, FL (originally published by IOP Publishing, London, U.K.)

Wu, F. and Qu, L., 2009, Diagnosis of subharmonic faults of large rotating machinery, *Mechanical Systems and Signal Processing*, 23, 467–475.

Yanussa-Kaltungo, A. and Sinha, J.K., 2014, Coherent composite HOS analysis of rotating machines with different support flexibilities, *Vibration Engineering and Technology of Machinery, Proceedings of VETOMAC-X*, Manchester, U.K., 9–11 September, Springer, New York, pp. 145–154.

9

Case Studies

9.1 Introduction

A collection of the principal tools available for the monitoring and fault diagnosis of vibration problems in rotating machinery is described in the earlier chapters of this book. Each investigation takes on its own characteristics, and there is no such thing as a standard investigation. There are nevertheless some standard approaches and, more importantly, disciplines to be adhered to. In this chapter, five differing studies are described in some details in order to convey an overview of the application of the techniques employed. Four of the studies report investigations successfully completed, whilst the fifth briefly describes an ongoing project related to a topic of concern in many countries at the present time.

9.2 Crack in a Large Alternator Rotor

The events reported in this section occurred many years ago, in 1981 at a time when measurement, data processing and computing facilities were less than current standards. Nevertheless, the diagnosis of the machine fault provides an admirable example of the use of vibration data in machine fault diagnosis. In May of that year, the machine returned to service following repair on the alternator which involved refitting of the outboard end-bell. These are heavy components which cover electrical connections on the rotor, and it is quite common for them to cause some imbalance problems following a refit. Experience shows, however, that there is a tendency to bed-in. On its return to service, the vibration levels on the run-up were remarkably consistent with measurements taken over a year before. There were, however, some changes in the response to loading.

The machine was a large turbo-alternator comprising a high-pressure (HP) turbine, intermediate-pressure turbine, three low-pressure (LP) turbines, the alternator and exciter. The vibration changes were most pronounced at bearing 11, the drive end of the alternator rotor.

The amplitude and phase of the synchronous and twice-per-revolution components for this bearing are shown in Figure 9.1. As shown in the figure, the vibration levels at that date were remarkably similar to those monitored 18 months earlier. It is not clear if the sensitivity to load was caused by a thermal process or whether it was an influence of the cracking which was subsequently diagnosed. The unit ran until 22nd July, during which time the synchronous vibration on bearing 11, after an initial increase, settled to the low level of 1.6 mm/s, whilst the twice per revolution showed a very slight gradual increase from 2.1 to 2.3 mm/s. The situation, however, was confusing, as these particular machines are known to exhibit load sensitivity in the 2/rev component.

The machine was then taken off load for refuelling and returned to service 8 days later. The vibration was plotted as a vector change from the values of the May run-up, and the results for bearing 11 are shown in Figure 9.2. There are some changes in the 2/rev, but they are very small and at this stage could not be regarded as diagnostic of a crack. There were, however, significant changes on the 1/rev components, suggesting a change in the state of balance. This could possibly be caused by settlement of the end-bell.

During August, there were changes in both the 1 and 2/rev vibrations. Whilst the 1/rev could possibly be a consequence of a crack, it was at least equally likely to be just part of the continuing settlement of the end-bell. The 2/rev showed an accelerating increase from 2.3 mm/s on 30th July to 3.3 mm/s on 28th August. All vibration levels, however, remained well within acceptable levels.

After a brief outage, the unit returned to service on 8th September, and Figure 9.3 shows the 1 and 2/rev at bearing 11 on this run-up, together with those in May and July. The 3/rev shown in Figure 9.4 has a very prominent peak at 660 rpm. This is close to one-third of the second critical speed of the generator rotor. Furthermore, the 2/rev shows a clear change at 1050 rpm, around half the critical speed. Taken together, these two factors were the first clear evidence of a crack in the rotor. Further evidence was observed in the 1/rev traces where a slight downward frequency shift was apparent in all the peaks indicating a reduction (albeit small) in the critical speeds.

The unit returned to high load, and on 10th September changes were observed on 1 and 2/rev, particularly on bearing 11. The 1/rev did not cause immediate concern as it was considered to be part of an ongoing trend, but the 2/rev increases appeared to confirm the provisional diagnosis of a crack. The levels continued to increase, and the unit was removed from service on 12th September when the 2/rev reached an agreed limit of 6 mm/s. The rundown is shown in Figure 9.5 where a number of features are apparent. All vibration components on the alternator bearings had approximately doubled in the 4 days of service since the outage. The changes in the 2/rev and 3/rev components that had, by this time, been taken to indicate a crack were more pronounced. Both generator critical speeds showed a reduction of

FIGURE 9.1
Comparison of bearing 11 vibration signatures, December 1979 and May 1981. (Data supplied by and reproduced with permission of EDF Energy, Gloucester, U.K.)

FIGURE 9.2
Plots showing changes as compared to the signature of May 1981. (Data supplied by and reproduced with permission of EDF Energy, Gloucester, U.K.)

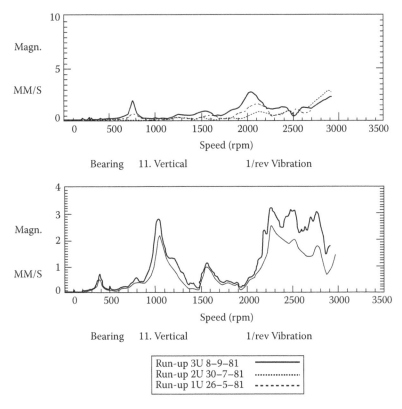

FIGURE 9.3
A comparison of run-up signatures on bearing 11. (Data supplied by and reproduced with permission of EDF Energy, Gloucester, U.K.)

around 5%, and the horizontal second generator critical speed recorded on bearing 12 had split into two peaks, at 1970 and 2060 rpm.

At this stage, other possible causes of the machine's behaviour were considered, but none were found to be fully compatible with the vibration behaviour. The diagnosis of a crack was confirmed, but the remaining issue was to identify its location in order to assist the maintenance staff in their investigation. Consideration of the rundown changes showed that the fault was on the generator or exciter rotors rather than in any of the turbine stages. The shift in the critical speeds of the generator rotor suggested that this is where the defect was since the much smaller exciter rotor would not have such a significant influence. On the generator rotor, the most significant changes are associated with the second critical speed; therefore, as discussed in Chapter 5, the defect is close to a point where the second derivative of the corresponding mode shape is a maximum. This leads to the conclusion that the crack is towards one end of the rotor.

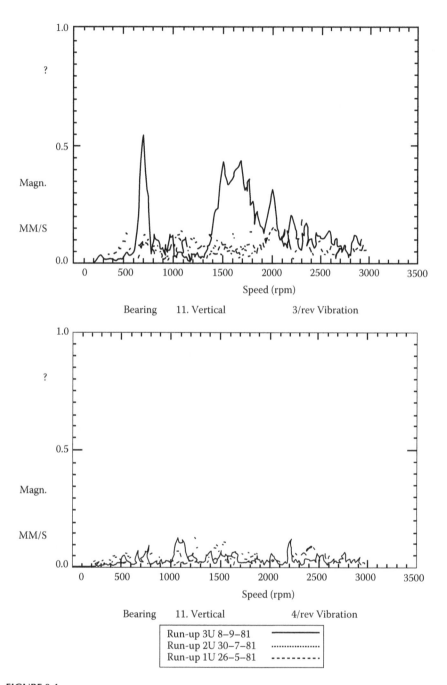

FIGURE 9.4
Higher harmonics at bearing 11. (Data supplied by and reproduced with permission of EDF Energy, Gloucester, U.K.)

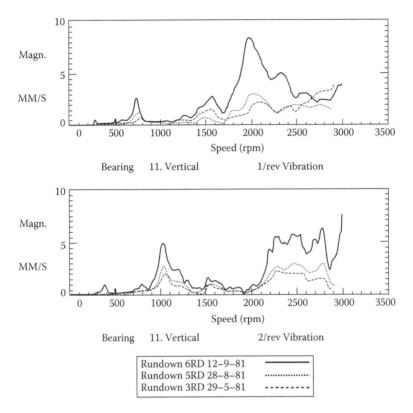

FIGURE 9.5
Bearing 11 rundown signatures. (Data supplied by and reproduced with permission of EDF Energy, Gloucester, U.K.)

When the rotor was removed from the machine, a large crack was found at the turbine end, close to a change in section. It extended to the bore and had an angle of around 140°.

The timely diagnosis had prevented a possible catastrophic failure.

9.3 Workshop Modal Testing of a Cracked Rotor

Over recent decades, many papers have been written on cracked rotor and this is appropriate, as it is an important practical problem, and the basic technology and phenomena are discussed in Chapter 5. The discussion here is of a technique often used to help identify and locate a crack, but one that needs careful attention when applied to electrical rotors such as generators.

With the suspected rotor out of the machine, a good means of crack detection and location is to perform a series of hammer impact tests with the rotor suspended from a crane. By repeating this test with the rotor at a series of orientations, the behaviour of the crack during operation can be simulated by gradually turning the rotor. The concept of this test is extremely simple: In orientations for which the crack is below the neutral axis, it will be at least partially open, whereas for other angles, the crack will close, and the vibrational properties of the rotor revert to those of the intact condition. The test may be enhanced by taking measurements at a series of axial locations, and the exercise then becomes a full modal test. Whilst this can be very informative, the focus here is the more limited objective of crack location, and this may be achieved with a single set of measurements.

For a rotor with axial symmetry, such as a turbine, at any angle the response to an impact will show a spectrum with a series of peaks corresponding to the free-free natural frequencies. The assumption of free-free conditions is, of course, an approximation but is readily checked and a correction applied if necessary. It is to be expected that at orientation for which the crack is in the upper portion of the rotor, it will be effectively intact, and the frequencies of the spectral peaks (see Figure 9.5) will be the resonances of the intact rotor. As the angle changes, however, the crack will open and the frequencies of all these resonances will reduce. The angle at which the resonant frequencies are at a minimum indicates the position at which the crack is fully open (assuming that there is a single crack), and, furthermore, the relative changes of the different peak frequencies between the intact rotor and those with the crack fully open can be used to locate the crack axially.

In Chapter 5, there is a brief discussion of the very elegant analysis by Mayes and Davies (1976) which related the change in natural frequency to the crack depth and location. It was shown that

$$\Delta\left(\omega_N^2\right) = R y_N''(s_c) \tag{9.1}$$

where $y_N''(s_c) = \left[\dfrac{d^2 y_N}{dz^2}\right]_{z=s_c}$, that is, the second derivative of the mode shape at the crack location s_c, and R is a function of crack and rotor geometry. y_N is the Nth mode shape and the rotor lies along the z axis. In Mayes and Davies (1984), the same authors extend their study using dimensional analysis to describe the stress concentration factor at the crack front. Their study leads to the simplified expression (shown in Chapter 5, repeated here for convenience)

$$\Delta\left(\omega_N^2\right) = -4\left(\frac{EI^2}{\pi r^3}\right)\left(1-v^2\right)F(\mu)y_N''(s_c) \tag{9.2}$$

where
　　μ is the non-dimensional crack depth (relative to the local shaft radius)
　　F is a function independent of all other parameters and is a universal function for a given shape of crack

The other parameters have the same meanings as in Chapter 5.

In principle, the function $F(\mu)$ can be derived from the appropriate stress concentration factor, if this is known. A different approach was taken in Mayes and Davies (1984) where the authors inferred the values of F from a series of experiments with chordal cracks. It was shown that to a very good approximation, this function is just equal to the fractional change in the second moment of area as measured at the crack face. Studies on cracks are summarised by Friswell et al. (2010).

In performing this calculation, however, it is important to note that the second moment of the remaining portion of the face is referenced to the new geometric centre rather than the original one. Hence, F is a very non-linear function of μ. Having established the change in stiffness and natural frequency arising from a crack, it is straightforward to represent these factors in a finite element (FE) model. In the same paper, it is shown that a crack may be represented by reducing the second moment of area of a single element by ΔI. Using a Rayleigh type of approach, it can be shown that if the second moment of area of the element concerned is I_0, then

$$\frac{\Delta I/I_0}{(1-\Delta I)/I_0} = \frac{r}{l_{el}}\left(1-v^2\right)F(\mu) \tag{9.3}$$

where l_{el} is the length of the section with reduced properties. Note that this parameter is at the discretion of the modeller within a reasonable range, but the choice determines the value of second moment of area. For any given chordal crack, there are two values of F and two orthogonal directions in the plane of the crack. Hence, the parameters of a representative model are now fully specified.

All this means that provided there is a reasonably accurate model of the rotor, and hence knowledge of the (un-cracked) mode shapes, the relative effect of the crack on each of the modes can be estimated. The significance of this is that it provides a crucial means of determining the location of a crack without completely dismantling the rotor. It is worth noting here that many industrial rotors are very complicated assemblies, and it is often very difficult to locate even a major crack.

Test procedures vary, but often eight equally spaced orientations will be used and the response to a hammer impact is recorded. The frequencies of the first two or three flexural resonances are noted in each position. It is sufficient to measure in a single direction, normally the horizontal, but as discussed in the following text, there are advantages in monitoring both orthogonal directions. With eight orientations, there are four pairs of diametrically opposite pairs for which differences in natural frequencies are measured, and it is these differences that lead to the crack identification. In its simplest terms, the following statements can be made:

1. The orientation at which the frequencies are minimum indicates and corresponds to the angular position at which the crack is at its lowest position and is consequently fully open.

2. The ratio of the frequency differences of the resonances locates the axial position of the crack.

3. Once the position of the crack has been located, the magnitude of the crack can be determined by evaluating the function F in Equation 9.2 and hence the non-dimensional crack depth μ.

Of course, this basic procedure may lead to partially contradictory conclusions, and it may be appropriate to use some form of least square procedure but this is not discussed here.

It is, however, worth considering some causes of confusion in executing this procedure. One important issue here is that whilst the crack is modelled as planar and chordal, this will at best be an approximation to reality. It must be emphasised that all cracks are different and the purpose of models is to extract key general features rather than accurately reproduce the behaviour of any specific case. Perhaps more importantly, the nature of the rotor itself must be considered: in particular, electrical rotors, such as generators and motors are not axisymmetric and this leads to considerable confusion.

An electrical rotor has different stiffness in two orthogonal directions, and this has some important consequences. First, it is important to appreciate that a crack may be recognised on an operational (horizontal) rotor by a change in the twice-per-revolution component of vibration. In an axisymmetric rotor the change will always be an increase, but this is not the case with a rotor which has an inherent degree of asymmetry. This is because the orientation of a crack may be such as to initially decrease the overall asymmetry which essentially drives the twice-per-revolution vibration: Of course, as a crack grows, the asymmetry will at some stage show an overall increase. The salient consideration is the vector change in the twice-per-revolution vibration. This will almost always be smaller than the synchronous vibration, but it is very 'stable' and in most machines shows very little variation making it a very useful diagnostic tool. Second, the introduction of a crack introduces cross coupling between the two principal planes, and this must be taken into account in interpreting the hammer tests described earlier. The cross coupling means, for instance, that when excited, predominantly in the 'weak' direction, it may well produce a displacement in the 'strong' axis, and due allowance must be made for this, complicating the diagnosis considerably.

Consider the example shown in Figure 9.6. A uniform rotor has a 5% difference in stiffness in its two principal directions and consequently a 2.5% difference in the free-free natural frequencies. A crack has been introduced at the midpoint which has a non-dimensional depth of $\mu = 1$, that is the crack effectively removes half of the total cross section, and the crack face tip is at an angle of 45° to the principal axes of the rotor. The situation was modelled using the reduction of second moments of area given by Equation 9.3, and the

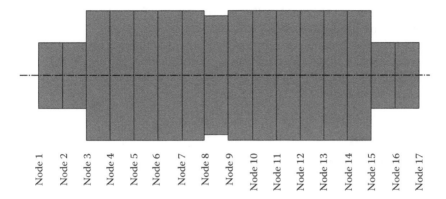

FIGURE 9.6
Rotor model to check cross coupling.

location is identified in Figure 9.6. The rotor is considered to be suspended from a crane, and hence there are no bearings in the model. This was considered by Lees and Friswell (2001).

With the rotor intact, there is no cross coupling of motion in the two directions of the principal axes, but this changes markedly with the crack. Figure 9.7 shows the maximum displacement in the orthogonal direction relative to the excited direction. In a number of modes, the degree of coupling

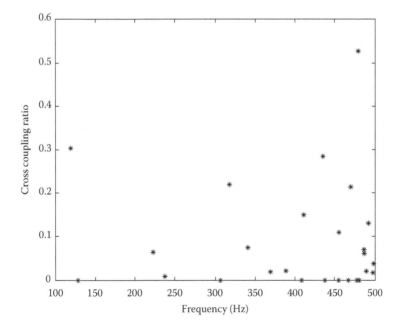

FIGURE 9.7
Cross coupling of directions in a cracked rotor.

is pronounced and this makes the interpretation of simple resonance tests somewhat more complicated.

Once this phenomenon is recognised, it is readily taken account of. Indeed, the sensitivity of this cross coupling is an interesting feature, which could possibly provide further diagnostic tools for crack detection.

9.4 Gearbox Problems on a Boiler Feed Pump

The machine in question was the main boiler feed pump at a large coal-fired power station. The main turbo-generators were rated at 500 MW, and each of the four units has a main boiler feed pump and two start and standby units. At the time of the investigation, the plant was relatively new but the main boiler feed pumps had suffered a number of failures and poor reliability. This was costly because although the turbo-generators could operate at full load using only the start and standby pumps, this mode of operation bypassed part of the feed heating system, and consequently efficiency was significantly reduced. The study was reported by Lees and Haines (1978).

The main boiler feed pump supplies 1.6 million kilogram of water per hour at a pressure of 220 bar and a temperature of 160°C. Clearly under these conditions, some preliminary pressurisation is needed to suppress cavitation at the impeller inlet, and for this reason, a small so-called booster pump is incorporated into the system. This booster pump is driven from the main pump via nominally 2:1 reduction gearbox (actually 111:56). This gearbox was the prime source of the problems encountered, with several gear failures, broken couplings and a cracked shaft. The whole arrangement is driven by a single-stage turbine rated at 17 MW with a maximum rotational speed of 5000 rev/min. Of this power, about 10% is transmitted to the booster pump. The turbine, HP pump, gearbox and booster pump are all connected by flexible couplings in order to allow for the substantial alignment changes with temperature during operation. A schematic of the complete arrangement is shown in Figure 9.8.

The overall problem was clearly to resolve the many and varied problems on the unit. At that time, there were four units and this power station, but another four at two other sites: all exhibited a similar pattern of faults. The challenge was to determine if there was a single underlying cause and thereby rectify the issue. A feature of this design of pumps was to use a balance drum together with an external bearing to balance the thrust, rather than the (at the time) more conventional balance-piston arrangement. The only relevance of this was that the design involved a long overhang from the bearing to the coupling, influencing the torsional properties. As the gearbox appeared to be the focus of problems, it was logical to attempt to

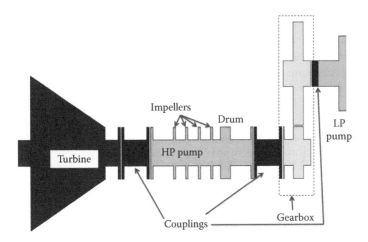

FIGURE 9.8
Idealised main boiler feed pump and turbine layout.

monitor the transmitted torque, despite the inherent difficulties in making this measurement.

The torque measurement was achieved by bonding strain gauges in the spacer shaft of the double-membrane coupling between the HP pump and the gearbox. The signals were transmitted to a stationary receiver by means of a telemetry system – and the observations were both surprising and illuminating. At that time, this proved to be a challenging instrumentation exercise and further details are given by Lees and Haines (1978). It proved, however, to be very worthwhile and led to a solution of this and some related problems, although initial measurement results were puzzling, requiring a mathematical model to interpret the observations.

The measurements showed significant fluctuations in torque over a wide range of pump speeds and duties. A most important observation was that under some operational conditions (during run-up to normal speed) the fluctuations were larger than the mean torque, the consequences of which will be discussed in the following text. The main objective was to ascertain the origin of these fluctuations. Another important observation was that over a range of speeds torque measurements revealed three distinct frequency components (albeit varying in magnitude and relative importance). The three components were the rotational frequencies of the two shafts together with a fixed frequency of about 39 Hz. It was later established that this fixed frequency term corresponds to the first torsional resonance of the system which is excited by non-linear terms in the equations of motion.

A finite element model of the system was initially developed based on the plant drawings, comprising 52 beam elements. As the investigation developed, however, it was observed that because the coupling between the turbine and the pump was relatively stiff, it was possible to work with the simplified model shown in Figure 5.17 comprising four inertias and two (torsional) springs. This is a purely torsional model: the neglect of coupling between torsion and lateral vibration was judged to be justified in the light of the plant measurements, although this effect can be included in the FE model.

Like any mechanical component, gears are not exact, but are constructed to within specified tolerances. Consider, therefore, the forces generated in the system resulting from a specified error on one of the gear components. The equations of motion for the simplified system are as follows:

$$I_1 \frac{d^2\theta_1}{dt^2} = K_1(\theta_2 - \theta_1) \tag{9.4}$$

$$I_2 \frac{d^2\theta_2}{dt^2} = K_1(\theta_1 - \theta_2) + T \tag{9.5}$$

$$I_3 \frac{d^2\theta_3}{dt^2} = K_2(\theta_4 - \theta_3) + \gamma T \tag{9.6}$$

$$I_4 \frac{d^2\theta_4}{dt^2} = K_2(\theta_3 - \theta_4) \tag{9.7}$$

where γ is the gear ratio. Before any analysis can be carried out, however, a constraint must be applied to model the way in which the gears transmit motion. T is the transmitted torque which is unknown *a priori*, but it must be sufficient to ensure that the gear components can move in such a way that the error profiles can be 'absorbed'. If the gearbox were perfect, an appropriate equation of constraint would be

$$\dot{\theta}_2 + \gamma \dot{\theta}_3 = 0 \tag{9.8}$$

However, since some errors are inevitable, the constraint is a little more complicated and may be expressed as

$$\frac{N_1}{2\pi} \dot{\theta}_2 + \frac{N_2}{2\pi} \dot{\theta}_3 = \varepsilon_1(\omega_2 + \dot{\theta}_2)\sin(\omega_2 t + \theta_2) + \varepsilon_2(\omega_3 + \dot{\theta}_3)\sin(\omega_3 t + \theta_3) \tag{9.9}$$

This is clearly a non-linear equation. Considerable simplification can be obtained by taking a case in which one of the two error terms dominate, say $\varepsilon_2 \gg \varepsilon_1$, and the torsional displacement is 'small' in some sense. Then multiplying through by $2\pi/N_1$ gives

$$\dot{\theta}_2 + \gamma \dot{\theta}_3 = \frac{2\pi\varepsilon_2}{N_1}\sin\omega_3 t \tag{9.10}$$

The system of equations may now be solved at each rotational speed. Chapter 5 shows how this can be achieved using general matrix constraints, but in this case, simple manipulation of the equations shows that

$$T = \dot{\theta}_2 \left[K_1 - I_2\omega^2 - \frac{K_1^2}{K_1 - I_1\omega^2} \right] = \alpha\dot{\theta}_2 \qquad (9.11)$$

And similarly

$$T = \frac{\dot{\theta}_3}{\gamma} \left[K_2 - I_3\omega^2 - \frac{K_2^2}{K_2 - I_4\omega^2} \right] = \frac{\beta\dot{\theta}_3}{\gamma} \qquad (9.12)$$

These expressions are now substituted into the equation of constraint (Equation 9.10) giving (after cancelling sinusoidal terms)

$$T \left[\frac{1}{\alpha} + \frac{\gamma^2}{\beta} \right] = \frac{2\pi\varepsilon_2}{N_2} \qquad (9.13)$$

Hence, the transmitted torque at frequency ω is

$$T(\omega) = \frac{2\pi\varepsilon_2}{N_2} \frac{\alpha\beta}{\alpha\gamma^2 + \beta} \qquad (9.14)$$

Clearly, this can be expressed explicitly, but the added complexity does not contribute to understanding. The essential element of this process is that it is displacement controlled. The important point to appreciate in this is that the magnitude of the torque oscillations is arising, and consequently the stresses in the gear teeth are dependent on the torsional stiffness of the shaft line – in particular the most flexible part, namely, the couplings. This is because, in essence, the process is displacement controlled. In contrast to many situations, the magnitude of the relative motions of the gear components is determined: the difference in the motion of pinion and wheel in the gearbox must be (at least) equal to the tooth error, and forces will be generated to ensure that this is the case. Assuming that the gears remain in contact, the relative motion will indeed be equal to the error profile prevailing at the time and point of contact.

There is an assumption here; the gear teeth have been assumed to be absolutely rigid and this not quite true. In reality, all gear teeth have some flexibility, and this can be easily incorporated into the model although it inevitably makes it a little more complicated. Values for tooth stiffness can be calculated for standard tooth profiles, and for this case, it was found

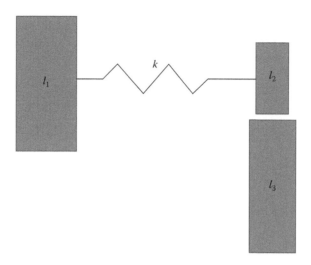

FIGURE 9.9
Simplified system for illustration of displacement control.

that the effect only becomes important for frequencies above 1 kHz and was, therefore, neglected.

Returning to Equation 9.14, consider the influence of the coupling stiffness values on the torque oscillations. The situation is already complicated, so to illustrate the effect of the displacement-controlled mechanism, consider now the much simpler situation shown in Figure 9.9 in which the driven component has simply inertia and the driving inertia is very large. For this simple system, the equations may be written as

$$I_1\ddot{\theta}_1 = k(\theta_2 - \theta_1) \tag{9.15}$$

$$I_2\ddot{\theta}_2 = k(\theta_1 - \theta_2) + R_2P \tag{9.16}$$

$$I_3\ddot{\theta}_3 = R_3P \tag{9.17}$$

P is the force of the gear teeth and other symbols retain their meanings as before.

Again damping terms have been neglected for the sake of simplicity but these will be included in the calculation of responses. If it now assumed that there is a single sinusoidal error on the pinion, then the equation of constraint becomes

$$R_2\theta_2 + R_3\theta_3 = e\sin(\omega t + \theta_2) \tag{9.18}$$

Note that e is a dimension on the gear rather than an angle. For this illustration, it is assumed that I_1 is very large and hence $\theta_1 = 0$. Then it follows that (neglecting second-order terms)

$$\left(k - I_2\omega^2\right)\theta_2 = R_2 P \tag{9.19}$$

Provided θ_2 is sufficiently small, Equation 9.18 can be linearised to give

$$\theta_2 + \gamma\theta_3 = \frac{e}{R_2}\sin\omega t \tag{9.20}$$

The consequences of this linearisation are discussed later, but it does make the solution to the problem much more tractable. Using the equation of constraint (9.20) in (9.17) yields

$$T\left[\frac{1}{k - I_1\omega^2} - \gamma^2\frac{1}{I_2\omega^2}\right] = \frac{e}{R_2}\sin\omega t \tag{9.21}$$

Simple manipulation now gives the transmitted torque oscillation due to a sinusoidal error of magnitude e as

$$T = \frac{I_2\omega^2}{\gamma^2}\frac{\left(k - I_1\omega^2\right)}{\left[k - \left(I_1 + \dfrac{I_2}{\gamma^2}\right)\omega^2\right]}\frac{e}{R_2}\sin\omega t \tag{9.22}$$

This result is fully consistent with the analysis of the more realistic system, but it clarifies the importance of coupling stiffness (or more precisely, shaft line stiffness in general) in determining the stresses and moments arising due to gear errors – and such errors are always present, but hopefully within limits set by various standards organisations (e.g. ASME, BS and ISO). Suppose for instance I_1 is small, with coupling stiffness K, then above the resonant speed, the torque fluctuation becomes

$$T = -\frac{ke}{R_2}\sin\omega t \tag{9.23}$$

Hence, the fluctuation is entirely determined by the stiffness. Since the process is displacement controlled, this should not be surprising; the gear component must move in such a way as to allow for the geometrical errors imposed, and hence if the shaft is very stiff (torsionally), high moments will be generated.

Returning to the more realistic (four inertia) model of the feed pump train calculations revealed that the system would generate high torque fluctuations with very modest errors on the gears. The gears themselves were well within standards, but the machine was just too sensitive to these errors. During machine run-up, the torque fluctuations exceed the mean torque on the machine, and hence the gears will lose contact and then suffer a significant impact as they re-engage. Although not conclusively proven, it was strongly suspected that this impacting caused the major damage to the gears.

The model was used to compare the plant with proposed solutions given a gear error of 25 µm on the slow slide gear wheel and the results are shown in Figure 9.10. Also shown in this figure is the mean torque transmitted through to the gearbox as a function of speed. The solution proposed was a reduced stiffness of the high-speed coupling, by installing a spacer with reduced wall thickness, together with the introduction of some torsional damping by the use of a rubber block coupling on the low-speed shaft. Note that in these calculations, some damping has been introduced; this is straightforward but was omitted from the derivation mentioned earlier for the sake of clarity. The modifications were carried out some years ago, and the result has been highly satisfactory with greatly improved plant availability and reliability. Two features in Figure 9.10 are worthy of note: on the modified system, the

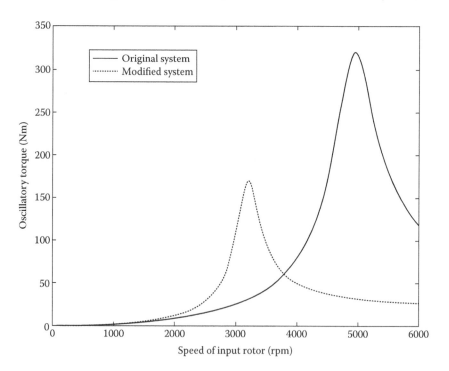

FIGURE 9.10
Torque oscillations due to wheel error of 25 µm.

peak level of induced oscillation is reduced, but it is also moved in frequency, away from the normal operational conditions. The four inertias are 15, 0.37, 3.38 and 5 kg m², respectively, whilst initially the two stiffnesses are 8×10^5 and 5.8×10^5 Nm/rad. In the modified system, the first stiffness is reduced to 3×10^5 Nm/rad.

One factor remains to be explained, namely, the presence of a component at a fixed frequency (39 Hz) over a wide speed range. This corresponds to the first torsional natural frequency and arises because of the non-linear term on the right-hand side of Equation 9.22. This is difficult to deal with, mathematically, but an analysis has been given recently by Lees et al. (2011) using modern computational methods. As reported by Lees and Haines (1978) insight may be gained by expressing the displacement as a series of harmonics of the shaft speeds (i.e., of both shafts). The non-linear term gives rise to products of the two series, which in turn leads to excitation over a wide range of frequencies. Given this broad-band spectrum, high excitation will occur at the (torsional) natural frequencies, a finding which is entirely consistent with the observed behaviour and the more modern analysis.

The modification, together with the introduction of damping in the second coupling, has been proved very successful. The problem highlights the importance of designing a complete system, rather than just an assembly of components.

9.5 Vibration of Large Centrifugal Fans

A rather simpler problem occurred on some exhauster fans of similar design at two coal-fired power stations. Exhauster fans extract a mixture of air and coal dust from the mills and blow it into the boiler. The fans pass about 100 t of coal per hour, and there are two of these fans on each of the mills and each of the 500 MW turbo-generators has four coal mills, whose purpose is to pulverise the coal. The problem with these fans was simply one of the high vibration levels was so acute that it limited the life of the impellers resulting in high maintenance costs and sometimes loss of load.

The general arrangement is depicted in Figure 9.11. The motor is connected to the shaft through a flexible coupling and is supported separately, but this is not of concern in the current problem. The fan impeller is a six-bladed disc of about 2 m in diameter and has a mass of 0.8 t. Since its working material is coal dust, it is not surprising that the blades of the impeller erode and consequently, an initially well-balanced impeller soon deteriorates. It was the rate of deterioration that was the major problem at this time. Ideally, the fans should be run for about 6 weeks, by which time efficiency would lost owing to the severe wear of the fan blades. This had always been a problem on these fans and so it was decided to review their design. The supplier had assured

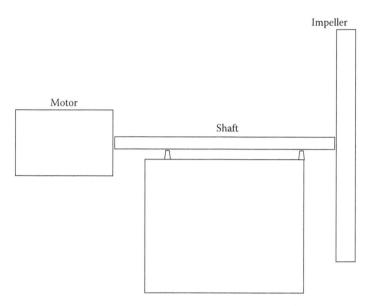

FIGURE 9.11
Exhauster fan.

us that the first critical speed was 1900 rpm and consequently running at 1500 rpm should be perfectly acceptable, but plainly this was not the case.

The supporting block was a fabricated steel structure which clearly has some flexibility, but initially it was considered to be rigid for the purpose of calculating the natural frequencies and critical speeds. The overhang of the impeller to the nearer bearing was 60 cm and the shaft diameter was 17 cm. The Campbell diagram is shown in Figure 7.6, and it is immediately obvious that there is a critical speed well within the fan's running speed range. These results caused some consternation with the suppliers. Their earlier assurances had neglected the effects of gyroscopic moments on the shaft. Whilst these effects are often relatively small for many rotors in which the major masses and inertias are within the bearing span, this is not the case with overhung rotors such as the case on these fans.

It is clear that the design of the fan required some revision. The parameters which could be changed relatively easily were the pedestal stiffness and the length of the overhang between the outer bearing and the impeller. Table 9.1 shows the critical speeds for a few combinations of these variables, and it is clear that shortening the overhang has an important part to play. However, whilst critical speeds are an important consideration, attention must be paid to the excitation. Noting that the predominant concern here is backward whirl modes, it is important to appreciate that reverse modes can only by excited by imbalance if there is asymmetry in the supporting structure.

To illustrate the point, consider the response to imbalance in the last two cases of the table. To calculate the response, however, some assumptions

TABLE 9.1

Variation in Critical Speeds

Case	K_x (N/m)	K_y (N/m)	Overhang (mm)	Critical Speeds (rpm)
A	5×10^8	5×10^8	600	−1245, +2250
B	1×10^8	5×10^8	600	−1178, +1945
C	5×10^8	5×10^8	300	−1680, +5200
D	1×10^8	5×10^8	300	−1600, +3000

must be made concerning the damping and this, as always, presents some difficulty. Clearly, there will be some damping in the bearings, but logically some damping would be expected from the fan impeller. Any precise evaluation of this influence is difficult, but fortunately, it is of little importance to the current discussion. Since the argument here is a comparison of different options, provided a consistent damping model is used, appropriate conclusions can be made. Figure 9.12 gives a comparison of the imbalance response of the four cases from Table 9.1. The peak around 1200 rpm in case B arises from a backward whirl, which disappears for the symmetric cases. This is because the troublesome reverse mode cannot be excited by imbalance forces if the supporting structure (including the bearings) is symmetric in the two directions orthogonal to the rotor. This point is discussed extensively

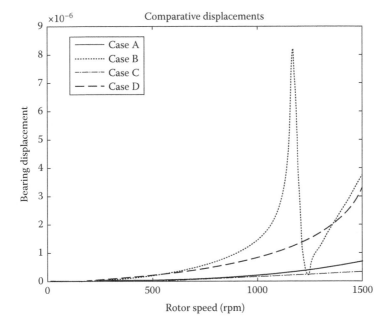

FIGURE 9.12
Response of options to nominal imbalance.

by Friswell et al. (2010), but it may be viewed simply as a consequence of the phase relationships between the displacements in the two directions. It should be emphasised that the critical speed still exists in case D as would become obvious if, for instance an external exciter was applied to the machine in a single direction.

Following a detailed study of the fans, two changes were implemented; the overhang was reduced to 300 mm (the minimum which could be accommodated without major design of the impeller and casing) and the fabricated support was replaced with a reinforced concrete block. Note that the objective of the latter was not an increase in stiffness but rather the achievement of approximately equal stiffness in the two orthogonal directions. Since the modifications, impellers have operated until decline in performance determines a change. Vibration is no longer an issue.

9.6 Low-Pressure Turbine Instabilities

Another interesting problem has arisen in the United Kingdom. A large power station comprising four 500 MW turbo-generators have been subject to blade cracking on the last stage of the LP turbines on to out of the four units. Measurements revealed an intermittent band of excitation of 7–12 Hz on the LP turbines which was not evident elsewhere on the machine. Furthermore, this low-frequency band modulates the synchronous vibration and its harmonics. A study of this particular problem is given by Hahn and Sinha (2014).

The interaction of the low-frequency band with the 1/rev vibration gives an excitation around 43 Hz, very close to a critical speed calculated as 46.88 Hz. Modulation with the 2/rev gives an excitation at 93 Hz, whilst the natural frequency of the final row blades is 90 Hz. There is an ongoing investigation into this problem, but early measurements, coupled with model studies, offer some insight into the sources of the difficulties.

The overall problem is complicated and believed to be due to a stall mechanism within the turbine. It occurs at low load, not surprising as under these circumstances, the turbine is operating well away from its design conditions. The flow of steam is, under these circumstances, not collinear with the rotating and static blades and, consequently, secondary flow patterns are initiated. This is entirely consistent with the fact that the two units displaying this problem are those which have had significant two shifting and operation at reduced load.

This is just one manifestation of a problem which is arising across the world. Rieger (2012) gives an overview of blading problems and discusses this stall phenomenon. Owczarek (2011) discusses in detail the physical basis of the problem. Although problems of stall have been known for a considerable time, concern has risen in recent years owing to two contributory and partially conflicting factors as follows:

1. In the drive to improve efficiency, blades tend to be free standing without the cover bands or lacing wires of earlier designs.
2. Fossil plants tend to two shifts or operate at reduced load as renewable energy takes up part of the demand.

Given that nuclear units must generate at high load for both technical and financial reasons, it is inevitable that greater flexibility will be required of fossil plants, and this presents a significant design challenge.

Hahn and Sinha (2014) have explored their data using conventional Fourier-based approach and these have proved effective. Wu and Qu (2009) report a related stall phenomenon on a blower unit using empirical mode decomposition (see Section 8.6.8). Their approach visually expresses the two-state behaviour.

9.7 Concluding Remarks

Several case studies are described in this chapter; some go back a number of years, whilst others show some current issues. All, however, relate to plant currently in service and hopefully give some insight into issues arising on real machines. Over recent decades, both instrumentation and data-handling facilities have seen dramatic developments. Analysis tools have also provided much greater prospect for the understanding of processes within machines. It is this understanding of the basic machine operation which remains at the heart of all condition monitoring and related disciplines.

Problems

9.1 An alternator rotor may be idealised as a uniform beam with mass 50 t and supported on bearings with stiffness 5×10^8 N/m spaced 6 m apart. Determine the first two natural frequencies.

9.2 The rotor of Problem 1 has a chordal crack 2 m from the first bearing. Calculate the change in the first two natural frequencies if the crack has non-dimensional depth of 0.25, 0.5 and 1.

9.3 There is an imbalance of 0.002 kg m at the first balance plane and the crack has non-dimensional crack of 1. Calculate the synchronous and twice-per-revolution components of vibration as functions of rotor speed. (Note: Chapter 5 has details of this type of analysis.)

9.4 A single stage gearbox has a gear ratio of 2.5:1. The pinion has moment of inertia of 100 kg m^2 with a radius of 100 mm, and the wheel has inertia 800 kg m^2. The pinion is connected to a motor of inertia 1000 kg m^2 through a coupling of stiffness 10^5 Nm/rad and the wheel is connected to a pump unit of inertia 500 kg m^2 through a coupling of stiffness 0.58 × 10^6 Nm/rad. If the wheel has a cyclic error of 10 μm, determine the magnitude of torque fluctuations as a function of input rotor speed. Assume that each coupling has damping of stiffness/2000.

9.5 The rotor of a fan has an overall length of 2 m and a diameter of 0.2 m. It is supported on two bearings with a constant stiffness of 4 × 10^5 N/m. The first bearing is 0.2 m from one end of the shaft which is driven via a flexible coupling. The second bearing is 0.5 m from the other end of the rotor. Overhung is an impeller with mass of 800 kg and diameter of 1.8 m.

Draw the Campbell diagram up to a rotor speed of 2000 rpm.

References

Davies, W.G.R. and Mayes, I.W., 1984, The vibrational behaviour of a multi-shaft, multi-bearing system in the presence of a propagating transverse crack, *Journal of Vibration, Acoustics, Stress and Reliability in Design*, 106, 146–153.

Friswell, M.I., Penny, J.E.T., Garvey, S.D. and Lees, A.W., 2010, *Dynamics of Rotating Machines*, Cambridge University Press, New York.

Hahn, W.J. and Sinha, J.K., 2014, Vibration behaviour of a turbo-generator set, in *Proceedings of VETOMAC-X*, Manchester, U.K., 9–11 September, pp. 155–161.

Lees, A.W. and Friswell, M.I., 2001, The vibration signature of chordal cracks in asymmetric rotors, in *19th International Modal Analysis Conference*, Kissimmee, FL.

Lees, A.W., Friswell, M.I. and Litak, G., 2011, Torsional vibration of machines with gear errors, in *9th International Conference on Damage Assessment of Structures (DAMAS)*, Oxford, U.K.

Lees, A.W. and Haines, K.A., 1978, Torsional vibrations of a boiler feed pump, *Transactions of the American Society of Mechanical Engineers, Journal of Mechanical Design*, 100(4), 637–643.

Mayes, I.W. and Davies, W.G.R., 1976, The behaviour of a rotating shaft system containing a transverse crack, *Institution of Mechanical Engineers Conference on Vibrations in Rotating Machinery*, Cambridge, U.K., pp. 53–64.

Mayes, I.W. and Davies, W.G.R., 1984, Analysis of the response of a multirotor-bearing system containing a transverse crack, *Journal of Vibration, Acoustics, Stress and Reliability in Design*, 106, 139–145.

Owczarek, J.A., 2011, On the phenomenon of pressure pulses reflecting between blades of adjacent rows in turbomachines, *American Society of Mechanical Engineers, Journal of Turbomachinery*, 133, 021012, 1–11.

Rieger, N.F., 2012, Progress with the solution of vibration problems in steam turbine blades, www.sti-tech.com/dl/vibnfr.pdf, Accessed 5 April 2015.

Wu, F. and Qu, L., 2009, Diagnosis of subharmonic faults of large rotating machinery based on EMD, *Mechanical Systems and Signal Processing*, 23, 467–475.

10

Overview and Outlook

10.1 Progress in Instrumentation

Recent years have seen significant advances in the instrumentation available for the monitoring and diagnosis of problems in rotating machinery. Over 50 years ago, the development of the non-contacting eddy current transducer was indeed a major step for the industry, and these are now a crucial component of many systems. More recent developments have been the refinements of wideband AE sensors which are enabling practitioners to progress some way beyond the basic process of counting events, the basis of early AE investigations. Other important developments have the enhancement of telemetry systems with significantly enhanced bandwidth capabilities.

Perhaps, more important is the development of wireless technology for instruments. Systems can now be controlled directly from laptop computers, and this offers significant potential for industry in general. In the past, cabling had been a very significant part of the cost overheads of monitoring systems, and hence the introduction of wireless communications has the potential to transform the economics of supervisory systems. It may, for instance become a viable option to monitor the behaviour of machines which were hitherto deemed uneconomic. In this context, it is worth emphasising that all effective progress in this field is ultimately governed by economics. For instance it is pointless analysing in detail the cause of a problem and means of alleviation if this cost is more than the replacement of the machine or component but these choices are, for the most part, fairly obvious.

10.2 Progress in Data Analysis and Handling

Of course, the real dramatic changes in recent years have been the availability of vast storage for copious volumes of data at minimal cost. This development has changed perspectives on the way in which plant data are stored, but it also presents challenges as to how best to sort, present and interrogate the databases. Hopefully, recent trends will continue for some time.

It will be recognised, however, that there is no direct link between volume of data and the information content, and this is a theme that has been stressed a number of times throughout the text. This is particularly apparent with regard to on-load plant data. The dilemma here is that there is a vast array of data available, but because the machine is running under quasi-steady conditions, the information content is limited. However, because of fluctuations introduced by the control system and external influences, there is some information to be gleaned, but only some forms of model, either statistically or physically based, can achieve this.

10.3 Progress in Modelling

The first numerical models of rotating machinery were based on the transfer matrix approach, a method which made very efficient use of the very limited computer storage available at the time. With the development of finite element techniques around 1960, this soon became the predominant method. The approach being linear is rather simpler than the transfer matrix method, albeit rather more demanding in terms of computer memory. The initial attraction of the transfer matrix approach was its modest storage demands but in recent years computer memory has become extremely cheap and so high storage requirements have ceased to be a limiting factor.

Early studies were undertaken with linear beam model and, as discussed in earlier chapters, these methods remain in use today. More sophisticated approaches using 3D elements are needed for the analysis of some rotors, but the appropriate methods are now well developed and documented but not discussed here. A discussion is given by Friswell et al. (2010). Some practitioners advocate the use of 3D modelling for all rotors, but the author disagrees with this view. The argument is that 3D models can be applied to all rotors regardless of profile. Whilst this is true, a very substantial proportion of rotors, probably a majority, are amenable to the simpler approach. The main benefit of this is dramatically faster execution times coupled with simple input and output. Underlying this is, perhaps, a difference in philosophy; some people aim to feed as much information as possible into a calculation and on occasions this has some merit. The other, perhaps more analytic, approach is to focus on a more basic calculation encapsulating the key physical components.

Finite element model (FEM) must now be regarded as a fully mature technology, and this represents a major achievement over the last five decades. Methods are available for the representation of many practical machines, and limitation is, in most cases, a lack of knowledge of internal details. An example of this is the treatment of rotor rub, as discussed in Chapter 6. There is no difficulty in deriving a theoretical model describing the process, but it is extremely difficult to determine the basic meaningful parameters needed to provide quantitative predictions. Two questions arise at this point, the first being what is possible and affordable? Second, and perhaps more important, what is required?

As discussed elsewhere, the fundamental goal in all condition monitoring activities must be economic. Whilst it may be academically satisfying to uncover details of the internal operation of a machine, there is little point if this does not lead to enhanced, faster or cheaper repairs. For minor plant items, a replacement policy is clearly an appropriate procedure.

Obviously, plant data are fundamental to progressing the understanding of machinery, yet of itself, the value is somewhat limited. It becomes invaluable, however, if it is used to gain insight into the internal processes of the machine, but there are a number of ways in which this can be achieved. This insight is achieved in three main routes, or, best of all, a combination of all three. The three contributory routes are as follows:

1. Detailed historical records of trends for each individual machine. At present, these records take the form of vibration amplitudes and phases during rundown. In addition, on-load data in the form of scatter plots form a valuable background. These should include records of any balancing procedures which may be utilised to save time in future operations.

2. Statistical models indicating the levels and variations in both on-load vibration levels and transient data. This will facilitate comparison with any changes observed in behaviour. These statistical models may extend to trained artificial neural networks (ANNs) to register the sensitivity of the system to significant operational parameters.

3. A mathematical model based on the physics of the machine's operation and the engineering drawings provide additional insight relating external measurement to external operation.

Of the three components described here, few would argue with the first two, yet there may be some practitioners who doubt the value of numerical models. It is true that with current understanding (and more often, limited data on clearances and other detailed parameters) global models often lack accuracy, but to some extent, this is of secondary importance. Although it is useful to have models that are a faithful reflection of the plant under operational conditions, the primary requisite is that the model can yield accurate sensitivities with respect to parameters.

ANNs are discussed in Chapter 8 and offer considerable scope in the understanding of machinery. These can be trained using a wide range of operational parameters and are easily adapted to determine the acceptability of a given machine state. Furthermore, an ANN can be used to categorise fault conditions, but this application, whilst useful, has some limitations. This is basically because the ANN is a statistical tool and has no physics built into it, unlike a formal mathematical model as outlined earlier. A consequence of this is that the ANN can only interpolate; it cannot extrapolate. In other words, the ANN can only classify the machine into a fault condition that is known to it (through previous data), whereas a mathematical model may be used to investigate new effects and new fault conditions.

This is not to decry the significant influence of neural computing in recent years. As classifiers of fault conditions, their application may be enhanced by using mathematical models to generate a wider range of pseudo-data for a variety of possible faults. This overcomes the perennial problem that operators have a vast array of data under normal operating conditions but relatively little in the presence of faults. Of course, the use of models always raises the issue of validation, but by recognising the importance of sensitivities rather than absolute accuracy, satisfactory qualification can be reached.

A further valuable application of ANNs is in determining the contribution of each of a series of contributory factors. This application was alluded to in Chapter 8 but not described in any detail: where the dependence on parameters is non-linear, the ANNs provide a much better insight than the linear methods of singular value decomposition.

When a network has been trained for a range of inputs, it is a straightforward and instructive exercise to perform a series of calculations on the network varying one input parameter at a time. In this way, the dependence of the system to each individual parameter may be investigated. Of course, it may be that cross dependencies between two or more input variables have an influence, but this too may be clarified by reference to the network. To date, this approach has found limited applications, yet it appears to have significant potential. The potential of this approach is discussed by Mayes (1994).

This leads to a larger, and very important, question of how to merge the power of neural computing with the predictive capabilities of a physical model, and here we can offer, as yet, no clear solution. Perhaps, the crucial distinction is that the physical model is constrained by the form of the equations: In Chapter 7, the fitting of a model to too many variables improves the fit to measured data but produces a model with limited predictive capability. In the same way perhaps, current ANNs have too much freedom, and although producing excellent data fitting, this is at the expense of predictive capability. This reasoning leads to the notion of some form of guided neural computing, but precisely what form this might take is unclear at present.

10.4 Expert Systems

A number of researchers have developed various 'expert system' concepts to the general problem of machine condition monitoring. The techniques used in these studies range from ANNs (Ben Ali et al. 2015), decision table analysis (Ebersbach and Peng 2008) and Bayesian methods (Xu 2012). The aim of these investigators is the encapsulation into software of the expertise of experienced assessors. Some convincing diagnosis has been shown for studies on rolling-element bearing faults. For some categories of machine, these systems offer a promising way forward, and evidence to date shows them capable of giving correct diagnoses. This will, at the very least, filter the more straightforward problems relieving pressure on the expert engineers.

Although the papers cited indicate promising capabilities, these logic-based systems give only diagnosis rather than any quantitative information as regards location or severity of a fault. This being so, the role of expert systems for large complex machines may be rather limited.

10.5 Future Prospects

There is currently a substantial array of tools available for the diagnosis of machinery ranging from instrumentation through data capture and storage to analysis and evaluation. It is worth considering the likely direction of trends in the foreseeable future, whilst recognising the dangers of prediction. In the previous section the distinctive powers of physically based models and neural method have been discussed and a long term objective must be to at least aim for the automatic diagnosis of machine faults. It is, of course, questionable whether this will ever be achieved, but any substantially reduction in the number of problems referred to specialist engineers would be a worthy objective. A range of possible developments now emerge and these are discussed in the following sub-sections.

10.5.1 Machine Diagnostics

The first of these relates to the optimisation of information of the internal condition of a machine. As discussed elsewhere, this inevitably relies on adequate instrumentation and recording of data but equally important is the interpretation of the measurements. This is where some form of model is needed, wherever possible a physics-based idealisation is preferred, but on occasions, statistical approaches can help. ANNs are certainly a help in checking various parameters against previous behaviour but are limited in extrapolating. However, there is a possible way around this difficulty.

A problem in the use of statistical approaches (including ANNs) to automatically detect and categorise fault condition is that there are insufficient data on faults; despite the continual occurrence of fault conditions, the proportion of available machine data under fault conditions is actually minimal.

Using plant models to artificially generate appropriate data can readily rectify this shortfall. To do this, the model used would need to be validated appropriately, but this is an example of complementarity between physical and statistical approaches to the area of machine diagnostics.

This begs the question of how far the technology can or should go. Any answer to this will be highly machine dependent. Clearly, a small machine that can easily (and cheaply) be replaced, cannot justify major expenditure of fault detection hardware and software. On the other hand, a large turbo-generator represents a major capital investment and loss of output is in itself very costly, and so the argument for effort towards fault detection and rectification is very significant. Several instances are discussed in other chapters which have illustrated the value of such methods. Models tuned to individual machines offer the prospect of much more sensitive tools coupled with the ability to infer more meaningful information from on-load data. Whilst the general trend of this is clear enough, any prediction of how far progress will reach is a matter of speculation. For high-value machines, it is not unreasonable to anticipate that a tuned model will run in parallel with the machine in real time being constantly updated. This would then provide a constant monitor with the capability to infer internal machine conditions and hence a rapid dissemination of faults.

Whether or not we will ever reach the ultimate point of truly automatic detection of faults remains a valid point. It seems to the author that this is perhaps a little doubtful, but really this is not a great concern. Provided diagnostics can reach a point to significantly reduce the number of cases needing referral to a human expert, then a significant objective will have been achieved. Furthermore, in the referred cases, the tuned model would provide the assessor with internal information not available at the current time without extensive investigative studies. Coupled to this would be the facility discussed in Chapter 7 whereby rather than merely saving vibration records, the refined system would record physically meaningful parameters of the machine, such as imbalance profile, bearing loads and the dynamic properties of the foundation structure. This in itself would significantly accelerate the resolution of problems.

To go beyond this takes us to the second strand of the vision, the concept of a 'smart machine'.

10.5.2 Self Correction ('Smart Machines')

Having developed and used models to determine the condition of an operating machine, the obvious next step is the possible application of forces to rectify a fault condition. This concept has been applied to structures with some success, but the application to machinery is in its infancy. Nevertheless, it is an area of considerable potential, and it is likely to gain attention of researchers in the foreseeable future. Whilst it is likely that early progress will focus on smaller machines, the effect will inevitably spread in due course.

Consider the ways in which an operating machine can be influenced by external factors; the possibilities are summarised in the following categories.

10.5.2.1 Magnetic Bearings

Magnetic bearings are, perhaps, the most developed of the technologies which may form part of future 'smart machines'. Here, the primary interest is in autonomous bearings in which the magnetic is controlled directly from the monitored shaft motion. Extensive work has been carried out on various aspects of their performance (see, e.g. Schweitzer 2009). Many are in operation on a very wide range of installations, ranging from small precision machines to large gas line compressors. They have many attractions but those cost high. A limitation to their usage is the limited load-carrying capacity which is substantially lower than that of a comparable oil journal bearing. It is worth noting that this limitation is really one of size: In principle, an active magnetic bearing (AMB) could be designed to take any load, but because of magnetic field saturation effects, capacity can only be increased by using larger conductors, and this reaches the point where further increases become impractical.

10.5.2.2 Electro-Rheological Bearings

At the end of the twentieth century, there was considerable interest in journal bearings that had particles of dielectric material immersed in the oil. Application of an electric field transforms the fluid from Newtonian to pseudo-plastic with very much higher viscosity. This is another route by which the application of an external influence modifies the machine's behaviour. A number of papers on this subject were published (see, e.g. Nikolakopoulos and Papadopoulos 1998) with promising results. However, little seems to have developed from these studies.

10.5.2.3 Externally Pressurised Bearings

This represents another established technology with some potential for smart machinery insofar as external forces can modify the bearing properties. As the name suggests, in these bearings, the pressure within the oil film is generated by an external pump rather than by the viscous flow within the journal. For this reason, these bearings are suitable for low-speed applications, but the added complexity does add to cost. The interest in the context of smart machine is that the external pump in effect modifies the bearing properties, and any components having this feature are of potential interest for the control of machines.

10.5.2.4 Other Approaches

Lees et al. (2007) designed a prototype bearing housing, whose stiffness could be varied by the application of a voltage. This was achieved by mounting the bearing in a collar of an elastomer material, the compression on which can be varied. Because the elastomer has highly non-linear properties, it becomes stiffer when it is compressed. The compression was imposed via two coils of shape memory alloy (SMA) which has the remarkable (and useful) property that shrinks when heated. The heating is achieved simply by passing a current through the coil. The discussion given in Lees et al. (2007) is at an exploratory stage, but it does raise the possibility of smart machinery, albeit for a fairly limited range of machines.

There are, however, a number of difficulties with this approach, an important one being the rate of response. Whilst there is some scope for applying this technology, difficulties arise in the cooling phase which demand some form of forced cooling for any useable response characteristic, and although this can be achieved, the complexity of the system increases. Although the response time is much slower than the rotational speed of the shaft, control is possible provided that the rate of speed change is slow.

Figure 10.1 shows the manner in which the maximum amplitude may be very significantly reduced during a machine run-up by varying the stiffness of the bearing support during the transient. Whilst this technology has some potential, SMA material performs most effectively in none cyclic

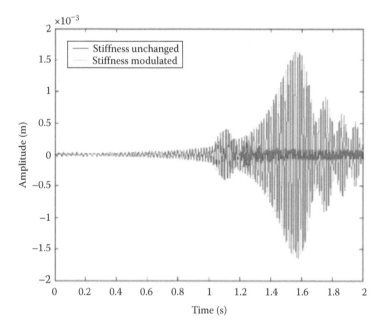

FIGURE 10.1
A controlled run-up. (Reproduced from Lees, A.W. et al., The control of bearing stiffness using shape memory, *Society of Experimental Mechanics, Proceedings of IMAC-XXV*, Orlando, FL, 2007. With permission.)

FIGURE 10.2
A 'smart' support structure. (Reproduced from Lees, A.W. et al., The control of bearing stiffness using shape memory, *Society of Experimental Mechanics, Proceedings of IMAC-XXV*, Orlando, FL, 2007. With permission.)

operations and can be sensitive to a range of parameters. The other difficulty is that modification of bearing stiffness will be a less effective means of control for machines with flexible rotors. Figure 10.2 shows the way in which a rolling-element bearing has been mounted surrounded by an elastomer whose non-linear properties are modified by forces exerted using the shape memory coils.

10.5.3 Shaft Modification

This is rather different to the three possible routes mentioned in the preceding section. Whereas the previous three actions exert influence via the bearings, here consideration is given to exerting a force, or combination of forces, directly on the shaft. This is important as it opens up the possibility of application to machines with flexible rotors. Whilst changes to bearing properties will always have some influence, if a rotor is flexible, the effect will be significantly reduced. Over recent decades, there has been a general trend towards the use of flexible rotors as this facilitates fast, smaller and more efficient installations. It is, therefore, highly desirable to develop technologies to promote smart machines, and so here consideration is given into the ways in which this might be achieved.

It makes sense to focus attention on the leading fault in all rotating machines, namely, rotor imbalance, discussed extensively in Chapter 4. Automatic balancing devices are currently in use on, for example washing machines (Rodrigues et al. 2011), but it is not clear that devices of this type could be applied more generally. There are, however, various ways in which the state of balance of a rotor could be altered during operation. One possible approach has been considered by Lees (2011) and Lees and Friswell (2014).

Both of these papers consider the same physical idea although the mathematical treatment differs slightly.

There are at least five possible approaches to use the forces arising in the piezoelectric MFC patches, which may be outlined as follows:

- Proportional feedback could be used in the rotating frame to directly change the natural frequencies. This would essentially change the natural frequency but it is unlikely to be applicable to large machines.

- Application of an axial force could, in principle, be used to modify the lateral natural frequencies. However, the axial force required is likely to be large and hence it is unlikely that this method would prove practical.

- A possibility is the use of derivative feedback to control damping rather than stiffness. This will improve the transient response and steady state operation if operated close to one of its critical speeds. There will be little effect across most of the operating speed range.

- Use the patches to induce rotor bends, which can counteract the unbalance. This is basically an application of standard balancing procedures. The piezo-electric patches are used to apply bending moments to the rotor and these replace the usual balance masses.

- Use integral feedback in the rotating frame. The unbalance response in the rotating frame is essentially a steady-state error, applying integral feedback can reduce this error and hence balance the machine.

The first three approaches are unlikely to be practical and hence this discussion will concentrate on the last two approaches.

Essentially the piezoelectric actuators apply moments to the shaft that may be easily controlled. The papers cited above (Lees, 2011; Lees and Friswell, 2014) both discuss this technique. The first paper treats the bend and the imbalance as separate terms whereas the later paper carries out a more direct calculation. The determination of the best moments to apply may be achieved either by means of a standard balancing procedure or by the use of a Kalman filtering identification approach. Methods of automatic balancing based on changing the mass distribution of the rotor do have the advantage that a balanced machine will be balanced at all speeds, although in practice the range of spin speeds where the balance is satisfactory is limited.

Integral feedback is well established to reduce the steady state errors in a wide range of control systems. This is an appropriate consideration since the imbalance on the rotor is a steady state parameter in the rotating frame. The error of interest in this case would be the response at the location of the sensors in the rotating frame. Note that this error would have components in both transverse directions and it is counteracted by two orthogonal bimorph actuators. Although the measurements of the vibration are taken from accelerometers and proximity probes in the stationary frame, the data is readily transformed into the rotating frame. A phase reference signal is an essential requirement, but this is now present on most machines.

Piezoelectric patches are small devices normally comprising lead zirconate titanate and some electrode connections. These are capacitive devices which exert a force when polarised by the application of a voltage. Originally used in the fine movement of components, these devices are now used extensively in 'smart structures'. They have several advantages over SMAs, the most important being a very much more rapid response and enhanced ability to work in a cycle as they are not dependent on thermal properties. The disadvantage is the high voltages required but this should be judged against the negligibly small currents demanded. They also have a rather smaller strain range than SMAs – no more than 2%, about one-third of which an SMA can tolerate.

The concept is to apply a bend to the rotor to compensate the imbalance. This is in direct contrast to the much more usual procedure of 'balancing out' a rotor bend. In the case studied by Lees and Friswell (2014), a shaft bend was formed by forces exerted by two pairs of piezoelectric patches mounted centrally on the rotor. Similar concepts have been studied by Horst and Wolfel (2004) and by Sloetjes and de Boer (2008). The work in this area shows promise but clearly significant development work is required in this area. The piezo-patches are activated by signals transmitted to the rotor by means of either slip rings or telemetry. A problem is that the voltages required by the patches are very high, although at negligible currents as they are capacitive devices.

After extensive discussion in Chapter 4, illustrating the important distinctions between a bend and imbalance, it is perhaps a little surprising that such an apparent equivalence is a possible route to smart machinery, but the important point is that the force imposing the bend is controllable and readily changed to suit rotor speed and conditions. As set out in Chapter 4, the big distinction of the two faults is the frequency variation. The studies carried out to date focus on the correction of a single mode, and whilst this is entirely understandable as a research investigation, it is likely that a real machine would need at least four patches per mode within the running speed range. More relevant, however, is a brief discussion on the connectivity of such a system, or how to get signals into the rotor.

Early studies have used piezoelectric MFC patches which are now widely used in smart structures where they have excellent high-frequency characteristics. Because of this, they are, perhaps, the natural choice for development. The high voltages (but minimal currents) can easily be input to the rotor by means of either slip rings or telemetry. These technologies are readily available but there would inevitably be some cost implication. An alternative strategy would be to transmit the signal at lower voltage with some amplification on the rotor. Again, this would appear feasible since the power involved is limited. There are various options that await a comprehensive investigation.

However, other possibilities offer significant potential. Rather than MFCs, the bending moments on the rotor could be applied using SMAs. The characteristics of SMAs are rather different to those of MFCs; the important differences are that they operate over a substantially greater range of strain (up to about 6% as compared with the typically 2% for MFCs) but have a much slower response, particularly in the cooling phase. A significant advantage, however, is a much lower voltage requirement (typically a few volts and against the 1.5 kV needed for some MFC activation), and this may be important in some applications.

The disadvantage of SMAs is the slower response time, but in many cases, this need not be a problem: although the vibration is relatively high frequency, the induced bend to bring about control is quasi-static. Therefore, the requirement for frequency response is governed by the rate of change of running speed rather than the speed itself, and this means that SMAs would be acceptable for some applications. Again, more research studies are needed here.

An intriguing possibility would be the use of magnetic shape memory alloys, a relatively new technology in which there have (at the time of writing) been few, if any, studies in their application to rotor systems. These have similar properties to SMAs in that phase changes give rise to geometric distortions but the changes are activated by magnetic fields rather than by temperature. The key benefit of this is that it raises the possibility of a non-contacting transmission to the rotor that may be rather simpler than telemetry. Of course, the potential benefits may be limited to certain types of machine.

The potential for transmission of signal to and from the rotor does raise other possibilities, for instance readings from strain gauges could yield direct alignment data and information on bearing performance. Of course, the extent to which these possibilities reach fruition is a matter of assessing the added value against the cost of implementation. Some basic machines will not justify the investment, but developments in the directions suggested are likely for some major plant areas.

The possibilities are very significant because if, using smart technologies, the significance of critical speeds and imbalance response can be negated or greatly reduced, then the stiffness and strength requirements of rotors can be reconsidered and design would be aimed at simply transmitting the required torque. This raises many and varied questions which hopefully will be addressed in the coming years.

It should be appreciated, however, that the ability to automatically apply correcting forces would represent a major step for rotating machinery and would completely change the constraints placed on designers. For instance the ability to compensate for imbalance would negate the need to avoid critical speeds in the running range. Often it is the need to avoid resonances (or critical speeds) which dictate the dimensions of rotors. Freed from the need to introduce rigidity, some rotors may become much lighter, needing only suitable dimensions to transmit the required torque. This could have great influence, as in many machines, the overall dimensions are chosen to avoid critical speeds rather than to convey the declared torque loading, and of course, this will have important repercussions on machine mass. Reduced mass will lead to reduced losses and higher efficiency, in addition to better usage of natural resources.

10.6 Summary

'Smart machines' are at the research stage and in the very early stages of development. However, they represent a culmination of work over recent decades in the inter-related disciplines of instrumentation, modelling and control.

The related fields of condition monitoring, rotor dynamics and control have seen significant strides in the last 50 years, and these have produced great economic benefits. The coming years should be equally productive.

References

Ben Ali, J., Fnaiech, N., Saidi, L., Chebel-Morello, B. and Fnaiech, F., 2015, Application of empirical mode decomposition and artificial neural network for automatic bearing fault diagnosis based on vibration signals, *Applied Acoustics*, 89, 16–27.

Ebersbach, S. and Peng, Z., 2008, Expert system development for vibration analysis in machine condition monitoring, *Expert Systems with Applications*, 34, 291–299.

Friswell, M.I., Penny, J.E.T., Garvey, S.D. and Lees, A.W., 2010, *Dynamics of Rotating Machines*, Cambridge University Press, New York.

Horst, H.-G. and Wolfel, H.P., 2004, Active vibration control of a high speed rotor using PZT patches on the shaft surface, *Journal of Intelligent Material Systems and Structures*, 15, 721–728.

Lees, A.W., 2011, Smart machines with flexible rotors, *Mechanical Systems and Signal Processing*, 25, 373–382.

Lees, A.W. and Friswell, M.I., 2014, Smart machines with flexible rotors, in *Proceedings of VETOMAC-X*, 9–11 September, Manchester, U.K., Springer, New York, pp. 855–864.

Lees, A.W., Jana, S., Inman, D.J. and Cartmell, M.P., 2007, The control of bearing stiffness using shape memory, in *Society of Experimental Mechanics, Proceedings of IMAC-XXV*, Orlando, FL.

Mayes, I.W., 1994, Use of neural networks for on-line vibration monitoring, *Proceedings of the Institution of Mechanical Engineers*, 208, 267–274.

Nikolakopoulos, P.G. and Papadopoulos, C.A., 1998, Controllable high speed journal bearings, lubricated with electro-rheological fluids. An analytical and experimental approach, *Tribology International*, 31(5), 225–234.

Rodrigues, D.J., Champneys, A.R., Friswell, M.I. and Wilson, R.E., 2011, Two-plane automatic balancing: A symmetry breaking analysis, *International Journal of Non-Linear Mechanics*, 46(9), 1139–1154.

Schweitzer, G., 2009, Applications and research topics for active magnetic bearings, in *Proceedings of the IUTAM Symposium on Emerging Trends in Rotor Dynamics*, Indian Institute of Technology, Delhi, India, 23–26 March.

Sloetjes, P.J. and de Boer, A., 2008, Vibration reduction and power generation with piezo-ceramic sheets mounted to a flexible shaft, *Journal of Intelligent Material Systems and Structures*, 19(1), 25–34.

Xu, B.G., 2012, Intelligent fault inference for rotating flexible rotors using baysian belief network, *Expert Systems with Applications*, 39, 816–822.

Solutions to Problems

Chapter 2

2.1 Maximum resolution sample over 71 rev = (initially 0.71 s) and the maximum resolution is

$$\Delta f = \frac{1}{0.71} = 1.4 \text{ Hz}$$

Overall sampling rate must be at least 800 Hz, block size 800 × 0.71 = 570 samples.

2.2 If the acquisition is fixed by rotor position, then the resolution follows Equation 2.14 and the resolution improved as the rotor decelerates. If it is at fixed time intervals, then the resolution stays fixed at its initial values assuming that the block size is constant.

2.3 Clearly, there are two important components:

- Case (a), time span 0.5 s, shows both components but with low resolution, and the 25 Hz component has suffered leakage.
- Case (b), time span 1 s, shows no leakage, both components accurately reflected.
- Case (c), time span 2 s, shows sampling interval now at 0.01 s and so cut-off frequency is 50 Hz, and so the higher-frequency component does not appear.
- Case (d), time span 4 s, shows that both components are aliased to zero owing to an inadequate sample rate.

2.4 The resolution must be at least 2.5 Hz preferably 1.25 Hz which means a sampling period of at least 0.8 s. The two frequencies are 25 and 27.5 Hz, and to avoid leakage, there must be an exact number of cycles of both signals; this suggests a 10 s sample, sampling at 55 Hz. But this gives a cut-off frequency of 27.5 and an aliased term at 30 which is confusing. So acquire at a higher frequency, say 100 Hz.

2.5 The time response can be expressed as

$$y(t) = \frac{1}{\Delta \omega t} \{\cos(\omega_0 t) \sin(\Delta \omega t)\} = \cos(\omega_0 t) \left(\frac{\sin \Delta \omega t}{\Delta \omega t} \right)$$

This can be plotted given the bandwidth and base frequency.

2.6 The standard deviations (semi-axes of the ellipse) are given by

$$\sigma_1 = \sqrt{38.14} = 6.18 \quad \sigma_2 = \sqrt{89.29} = 9.45$$

The angle of axis orientation is given by $\theta = \dfrac{180}{\pi} \times \cos^{-1}(0.5544) = -56°$. Probability is 0.2%.

Chapter 3

3.1 Natural frequencies are 17.51, 15.76 and 13.72 Hz.

3.2 First with four elements, $F = 32.3246 \ 105.1589$ Hz, and with eight elements, $F2 = 32.3213 \ 105.0695$ Hz.

3.3 Using Timoshenko elements, $F = 32.3114 \ 105.0193$ Hz, and with eight elements, $F2 = 32.3079 \ 104.9252$ Hz, that is very little difference as the shaft is slender.

3.4 For simply supported rotor $k = \dfrac{48EI}{L^3}$

- Masses of discs are 0.1286, 0.5881 and 1.3539.
- Mass of the rotor 3.6757 kg.
- Natural frequencies are 17.01, 14.43 and 12.08 Hz.

3.5 The Campbell diagram shows that the critical speeds are 2402, 8107 and 9684 rev/min.

Chapter 4

4.1 17.66 g at 200 mm and an angle of 148°.

4.2 Balancing with a bend

In this case, the parameters are $W = 22$ g at 200 mm radius and 164°. Note, however, that at the balance speed, the vibration will now be higher but will never exceed the magnitude of the bend.

4.3 The result for the two imbalances is 80 g at 200 mm on disc 1, at 358°, 168 g at 200 mm on disc 2, angle 84°. Hence, the correction masses will be applied 180° opposite in phase.

4.4 Modal balancing is explained, with an example in Section 4.3.2.

The required correction masses are as follows:

- On disc 1: 0.0103 kg m at 167°
- On disc 2: 0.0047 kg m at 53°

4.5 Required correction masses 0.0103 kg m at 347° on disc 1, and 0.0047 kg m at 233° on disc 2.

Chapter 5

5.1 Large peak at 550 rpm of 120 μm, second lower peak at 2350 rpm of 60 μm.

5.2 0.17 min = 5×10^{-4} rad, so first bearing must be raised by $2 \times 0.5 * 5 \times 10^{-4}$ m = 0.5 mm.

Second bearing must be raised by $2 \times 4.5 * 5 \times 10^{-4}$ m = 4.5 mm.

To ensure the outer bearings are at the same height, rotate the whole arrangement by 5×10^{-4} rad.

Brg 1: 0, Brg 2: $-4 \times 5 \times 10^{-4}$ = -2 mm, Brg 3: -2 mm, Brg 4: 0.

5.3 Moment is 1170 N m.

5.4 This is just a straightforward two (torsional) mass problems. The analysis follows that in the text, the rigid-body mode is not relevant to the problem. Torsional natural frequencies are 13.6 and 21.4 Hz. The peak response to the errors is 700 Nm.

5.5 The analysis of this problem requires the use of asymmetric rotor elements and the changes to the cross section as given by Equation 5.30. Frequencies of intact rotor 77.95, 31.69 and 70.92 Hz

Crack μ = 0.25 frequencies split 7.92, 7.93, 31.59, 31.6, 70.877, 70.879 Hz

Crack μ = 0.5 frequencies 7.88, 7.9, 31.39, 31.49, 70.79, 70.82 Hz

Crack μ = 0.75 frequencies are 7.79, 7.88, 30.98, 31.36, 70.61, 70.77 Hz

Crack μ = 1, Frequencies are 7.51, 7.87, 29.84, 31.34, 70.05, 70.77 Hz.

Chapter 6

6.1 This is readily solved from Equation 6.2. However, a numerical solution is required as it is difficult to deal with analytically. With viscosity 0.04 Pa s, eccentricity = 0.41. For e = 75% a viscosity of about 0.006 Pa s would be required which is a very low value. This is best solved by repeated solutions of Equation 6.2.

6.2 This is a straightforward application of the modelling software. Of course, a variety of calculation routines may be used following the logic in the main text. With original clearances the peak response is 1 mm at 1490 rev/min, but with increased clearances the peak response is 2 mm at around 855 rev/min.

6.3 At 5000 rpm, eigenvalues are $1.0e + 02$ * $-0.0065 - 2.8165i$, $-0.0065 + 2.8165i$, $-0.1800 - 2.8838i$, $-0.1800 + 2.8838i$, $-0.0605 - 7.8751i$, $-0.0605 + 7.8751i$, $-0.1239 - 8.1285i$ and $-0.1239 + 8.1285i$.

6.4 With reduced pressure, this becomes (rad/s) $0.53 - 241.92i$, $0.53 + 241.92i$. Note that these two are unstable! $-13.72 - 248.64i$, $-13.72 + 248.64i$, $-4.24 - 774.42i$, $-4.24 + 774.42i$ $-8.80 - 799.73i$ and $-880 + 799.73i$.

6.5 To answer this problem, we use Equations 6.24 through 6.29.

At 50% cutoff, $f = 38$ Hz; at 75%, $f = 71.3$ Hz; and at 90%, $f = 81.9$ Hz.

Chapter 7

7.1 The results are

$$k = 10^6 \begin{bmatrix} 1.85 & -0.72 \\ -0.56 & 0.79 \end{bmatrix} \quad M = \begin{bmatrix} 8.35 & -0.29 \\ -0.29 & 6.43 \end{bmatrix}$$

$$C = \begin{bmatrix} 408 & 118 \\ 118 & 440 \end{bmatrix}$$

These are not precise answers as the truncation of the input (to two decimal places and phase to the nearest degree) has introduced significant noise.

7.2 Revised estimates are

$K = [3.3 -1.33; -1.33 \ 0.9] \times 10^6$, $M = [16.3 -1.7; -1.7 \ 52]$, $C = [989 \ 49; 49 \ 323]$.

These are in substantial error, because not enough of the frequency range has been considered. In particular, only the first resonance is included.

7.3 The estimates with a diagonal damping matrix are

$K = [1.85 -0.72; -072 \ 0.79] \times 10^6$, $M = [8.35 -0.29; -0.29 \ 6.43]$.
$C = [488 \ 0; 0 \ 439]$.

7.4 The minimum eigenvalue for the regression matrix is as follows:

$-0.0000 \ -0.0000 \ 0.0000 \ 0.0004 \ 0.0019 \ 0.0068 \ 0.0307 \ 0.1950 \ 0.2808 \ 0.3188$. Note that the significant increase is at the second resonance.

This is reasonable even though it slightly distorts the original system.

7.5 Loads are 1608, 1114, 1670 and 1161 N, respectively.

Chapter 8

8.1 Ten micron limit is taken up at 1850 rpm.

8.2 Sampling period = 0.49 s. This means that the resolution is about 2 Hz. Initially, this will involve 500/60 * 0.49 rev, about 4.5 rev and if we need up to the fourth harmonic for monitoring, this implies a block size of 8 × 4.5 = 36 samples. This would be adequate above 120 rpm. Alternatively, use Vold–Kalman.

8.3 A slow rundown shows peaks at 850 and 2450 rpm.

8.4 See solution manual for rundown analyses.

8.5 Maximum velocity is 113.4 m/s, occurring at 22.2 s and the instantaneous frequency is 45.36 Hz.

Chapter 9

9.1 The two natural frequencies are 21.4 and 38.2 Hz.

9.2 Cracks give the following frequencies for non-dimensional crack depths of

0	0.25	0.5	0.75	1	
42.4624	41.9213	40.8130	38.7215	33.5258	mode 1 x
42.4624	41.9369	41.1645	40.6617	40.5705	mode 1 y
101.0152	100.4137	99.2497	97.2694	93.2518	mode 2 x
101.0152	100.4307	99.6095	99.0974	99.0063	mode 2 y
214.9278	213.9426	211.9073	207.9549	196.8865	mode 3 x
214.9278	213.9710	212.5561	211.6270	211.4578	mode 3 y

All frequencies in Hz.

9.3 The once-per-revolution excitation is determined in the usual manner. Twice per rev has a peak at half the critical speed, where amplitude rivals that of the synchronous signal.

9.4 Torsional resonances are 1.16 and 6.89 Hz. Peak fluctuation is 6 N m at 200 rev/min of input shaft.

9.5 First critical speed is 1299 rev/min which is a backward mode.

Index